A MANUAL

OF

PHOTOGRAPHIC CHEMISTRY.

A MANUAL

OF

PHOTOGRAPHIC CHEMISTRY,

INCLUDING THE

PRACTICE OF THE COLLODION PROCESS.

BY

T. FREDERICK HARDWICH,

LECTURER ON PHOTOGRAPHY IN KING'S COLLEGE, LONDON;
LATE DEMONSTRATOR OF CHEMISTRY IN KING'S COLLEGE.

Fourth Edition.

LONDON:
JOHN CHURCHILL, NEW BURLINGTON STREET.
MDCCCLVII.

PRINTED BY

JOHN EDWARD TAYLOR, LITTLE QUEEN STREET,

LINCOLN'S INN FIELDS.

PREFACE TO THE THIRD EDITION.

It is a source of much gratification to the Author to find himself called upon to prepare a Third Edition of his Manual in less than fourteen months from the date of its first publication. No greater proof could have been afforded of the rapid advance which the Photographic Art is now making in this country.

On once more entering upon the task of revision, the Writer has been led to reflect in what way the utility of the Work may be promoted; and from numerous inquiries he believes that this result will best be attained by carefully omitting everything which does not possess *practical* as well as scientific interest. The majority of Photographers look to the Art to furnish them with amusement as well as instruction, and they are deterred from entering upon a study which seems to involve a great amount of technical detail: these remarks however are not intended to discourage a habit of perseverance and careful observation, but simply to distinguish between the essential and the non-essential in the theory of the subject.

The present Edition differs in many important particulars from those which have preceded it. It has undergone a fresh arrangement throughout. In some parts it is condensed, in others enlarged. The Chapters on Photographic Printing are entirely re-written, and include the whole of the Author's investigations, as published in the Society's Journal. The minute directions given in this part of the Work will show how much success in Photography is thought to depend upon a careful attention to minor particulars.

Another point which has been kept in view, is to recommend, as far as possible, the employment of chemical agents which are used in medicine and vended by all druggists throughout the united kingdom. It is often an advantage to the Amateur to be able to purchase his materials near at hand; and, if the common impurities of the commercial articles are pointed out, and directions given for their removal, the 'London Pharmacopœia' will be found to include almost all the chemicals necessary for the practice of the Art.

Great additions have been made to the Index of the present Edition, which is now so complete that a reference to it will at once point out the most important facts relating to each subject, and the different parts of the Work at which they are described.

In conclusion, a hope is expressed that this 'Manual of Photographic Chemistry' may be found to be a complete and trustworthy guide on every point connected with the theory and practice of the Collodion process.

London, June 2nd, 1856.

PREFACE TO THE FOURTH EDITION.

THE Author has endeavoured to keep pace with the improvements which are daily being introduced in the science and art of Photography. In the present Edition alterations have been made in the style and general arrangement of the work, and additional matter has been inserted.

Since the publication of the Third Edition, a series of experiments have been made on the manufacture of Collodion, the results of which have thrown further light upon the conditions affecting the sensitiveness of the excited film, and have enabled the writer to introduce an organic substance, "Glycyrrhizine," which will be found of service in making Photographic copies of Engravings and similar works of Art.

Dr. Norris, of Birmingham, has within the last few months communicated a paper on *dry Collodion*, which places the theory of that subject upon a better footing than before. The Oxymel preservative process is now also thoroughly understood, and may be considered certain.

In addition to the above, the "Albuminized Collodion" of M. Taupenot, which experience proves to be one of the best dry processes at present known, is included in this Edition.

King's College, London, April 6th, 1857.

ERRATA.

Page 24, line 5, *for* conditions *read* condition.
Page 115, line 32, *for* Iodide *read* Iodine.
Page 194, line 15, *for* p. 88 *read* p. 188.

CONTENTS.

———◆———

PART I.

THE SCIENCE OF PHOTOGRAPHY.

CHAPTER VI.

THE PHOTOGRAPHIC PROPERTIES OF IODIDE OF SILVER UPON COLLODION.

CHAPTER VII.

ON POSITIVE AND NEGATIVE COLLODION PHOTOGRAPHS.

CHAPTER VIII.

ON THE THEORY OF POSITIVE PRINTING.

CHAPTER IX.

ON THE DAGUERREOTYPE AND TALBOTYPE PROCESSES.

PART II.

PRACTICAL DETAILS OF THE COLLODION PROCESS.

CHAPTER I.

PREPARATION OF COLLODION.

CHAPTER II.

FORMULÆ FOR SOLUTIONS REQUIRED FOR COLLODION PHOTOGRAPHS.

CHAPTER III.

MANIPULATIONS OF THE COLLODION PROCESS.

CHAPTER IV.

THE DETAILS OF PHOTOGRAPHIC PRINTING.

CHAPTER V.

CLASSIFICATION OF CAUSES OF FAILURE IN THE COLLODION PROCESS.

CHAPTER VI.

LANDSCAPE PHOTOGRAPHY BY THE COLLODION PRESERVATIVE AND COLLODIO-ALBUMEN PROCESSES.

PART III.

OUTLINES OF GENERAL CHEMISTRY.

CHAPTER I.

THE CHEMICAL ELEMENTS AND THEIR COMBINATIONS.

CHAPTER II.

APPENDIX.

A MANUAL

OF

PHOTOGRAPHIC CHEMISTRY.

———◆———

INTRODUCTION.

In attempting to impart knowledge on any subject, it is
not sufficient that the writer should himself be acquainted
with that which he professes to teach. Even supposing
such to be the case, yet much of the success of his effort
must depend upon the manner in which the information is
conveyed; for as, on the one hand, a system of extreme
brevity always fails of its object, so, on the other, a mere
compilation of facts imperfectly explained tends only to
confuse the reader.

A middle course between these extremes is perhaps the
best to adopt; that is, to make selection of certain funda-
mental points, and to explain them with some minuteness,
leaving others of less importance to be dealt with in a more
summary manner, or to be altogether omitted.

But independently of observations of this kind, which
apply to educational instruction in general, it may be remarked, that there are sometimes difficulties of a more
formidable description to be overcome. For instance, in-
treating of any science, such as that of Photography,
which may be said to be comparatively new and unex-

B

plored, there is great danger of erroneously attributing
effects to their wrong causes! Perhaps none but he who
has himself worked in the laboratory can estimate this
point in its proper light. In an experiment where the
quantities of material acted upon are infinitesimally small,
and the chemical changes involved of a most refined and
subtle description, it is soon discovered that the slightest
variation in the usual conditions will suffice to alter the
result.

Nevertheless Photography is truly *a science*, governed
by fixed laws ; and hence, as our knowledge increases, we
may fairly hope that uncertainty will cease, and the same
precision at length be attained as that with which chemical
operations are usually performed.

The intention of the author in writing this work, is to
impart a thorough knowledge of what may be termed the
" First Principles of Photography," that the amateur may
arm himself with a theoretical acquaintance with the subject before proceeding to the practice of it. To assist this
object, care will be taken to avoid needless complexity in
the formulæ, and all ingredients will be omitted which are
not proved to be of service.

The impurities of chemicals will be pointed out as far as
possible, and special directions given for their removal.

Amongst the variety of Photographic processes devised,
those only will be selected which are correct on theoretical
grounds, and are found in practice to succeed.

As the work is addressed to one supposed to be unacquainted both with Chemistry and Photography, pains will
be taken to avoid the employment of all technical terms
of which an explanation has not previously been given.

A SKETCH OF THE MAIN DIVISIONS TO BE ADOPTED, WITH
THE PRINCIPAL SUBJECT-MATTER OF EACH.

The title given to the Work is " A Manual of Photographic Chemistry," and it is proposed to include in it a

familiar explanation of the nature of the various chemical agents employed in the Art of Photography, with the rationale of the manner in which they are thought to act.

The division adopted is threefold :—

PART I. enters minutely into the *theory* of Photographic processes; PART II. treats of the *practice* of Photography upon Collodion; PART III. embraces a simple statement of the main laws of Chemistry, with the principal properties of the various substances, elementary or compound, which are employed by Photographers.

PART I., or "the Science of Photography," includes a full description of the chemical action of Light upon the Salts of Silver, with its application to artistic purposes; all mention of manipulatory details, and of quantities of ingredients, being, as a rule, omitted.

In this division of the Work will be found nine Chapters, the contents of which are as follows :—

Chapter I. is a sketch of the history of Photography, intended to convey a general notion of the origin and progress of the Art, without dwelling on minute particulars.

Chapter II. describes the Chemistry of the Salts of Silver employed by Photographers; their preparation and properties; the phenomena of the action of Light upon them, with experiments illustrating it.

Chapter III. leads us on to the formation of *an invisible image* upon a sensitive surface, with the development or bringing out to view of the same by means of chemical re-agents. This point, being of elementary importance, is described carefully ;—the reduction of metallic oxides, the properties of the bodies employed to reduce, and the hypotheses which have been entertained on the nature of the Light's action, are all minutely explained.

Chapter IV. treats of the fixing of Photographic impressions, in order to render them indestructible by diffused light.

Chapter V. contains a sketch of the *Optics* of Photo-

graphy—the decomposition of white Light into its elementary rays, the Photographic properties of the different colours, the refraction of Light, and construction of Lenses.
In the last Section of the same Chapter will be found a
short sketch of the history and use of the Stereoscope.

Chapter VI. embraces a more minute description of the
sensitive Photographic processes upon Collodion.   In it is
explained the chemistry of Pyroxyline, with its solution
in Alcoholized Ether, or *Collodion;* also the Photographic
properties of Iodide of Silver upon Collodion, with the
causes which affect its sensitiveness to Light, and the action of the developing solutions in bringing out the image.

Chapter VII. continues the same subject, describing the
classification of Collodion Photographs as Positives and
Negatives, with the distinctive peculiarities of each.

Chapter VIII. contains the theory of the production of
Positive Photographs upon paper. In this Chapter will be
found an explanation of the somewhat complex chemical
changes involved in printing Positives, with the precautions which are required to ensure the permanency of the
proofs.

Chapter IX. is supplementary to the others, and a brief
notice of it will suffice. It explains the theory of the Photographic processes of Daguerre and Talbot; especially
noticing those points in which they may be contrasted with
Photography upon Collodion, but omitting all description
of manipulatory details, which if included would extend
the Work beyond its proposed limits.

The title of the second principal division of the Work,
viz. "The practice of Photography upon Collodion," explains itself.   Attention however may be invited to the
fifth Chapter, in which a classification is given of the principal imperfections in Photographs, with short directions
for their removal; and to Chapter VI., which describes the
preservation of the sensitiveness of Collodion plates and
the mode of operating upon films of Albumenized Collodion.

In Part III. will be found, in addition to a statement of the laws of chemical combination, etc., a list of Photographic chemicals, alphabetically arranged, including their preparation and properties as far as required for their employment in the Art.

The reader will at once gather from this sketch of the contents of the volume before him, that whilst the general theory of every Photographic process is described, with the preparation and properties of the chemicals employed, minute directions in the minor points of manipulation are restricted to Photography upon Collodion, that branch of the Art being the one to which the time and attention of the author have been especially directed. Collodion is allowed by all to be the best vehicle for the sensitive Silver Salts which is at present known, and successful results can be obtained with a very small expenditure of time and trouble, if the solutions employed in the process are prepared in a state of purity.

CHAPTER I.

HISTORICAL SKETCH OF PHOTOGRAPHY.

THE Art of Photography, which has now attained such perfection, and has become so popular amongst all classes, is one of comparatively recent introduction.

The word Photography means literally "writing by means of Light;" and it includes all processes by which any kind of picture can be obtained by the chemical agency of Light, without reference to the nature of the sensitive surface upon which it acts.

The philosophers of antiquity, although chemical changes due to the influence of Light were continually passing before their eyes, do not appear to have directed their attention to them. Some of the *Alchemists*, indeed noticed the fact that a substance which they termed "Horn Silver," which was probably a Chloride of Silver which had undergone fusion, became *blackened* by exposure to Light; but their ideas on such subjects being of the most erroneous nature, nothing resulted from the discovery.

The first philosophical examination of the decomposing action of Light upon compounds containing Silver was made by the illustrious Scheele, no longer than three-quarters of a century ago, viz. in 1777. It was also remarked by him that some of the coloured rays of Light were peculiarly active in promoting the change.

Earliest application of these facts to purposes of Art.— The first attempts to render the blackening of Silver Salts by Light available for artistic purposes were made by Wedgwood and Davy about A.D. 1802. A sheet of white paper or of white leather was saturated with a solution of Nitrate of Silver, and the *shadow* of the figure intended to be copied projected upon it. Under these circumstances the part on which the shadow fell remained white, whilst the surrounding exposed parts gradually darkened under the influence of the sun's rays.

Unfortunately these and similar experiments, which appeared at the outset to promise well, were checked by the experimentalists being unable to discover any means of fixing the pictures, so as to render them indestructible by diffused Light. The unchanged Silver Salt being permitted to remain in the white portions of the paper, naturally caused the proofs to blacken in every part, unless carefully preserved in the dark.

Introduction of the Camera Obscura, and other Improvements in Photography.—The "Camera Obscura," or darkened chamber, by means of which a luminous image of an object may be formed, was invented by Baptista Porta, of Padua ; but the preparations employed by Wedgwood were not sufficiently sensitive to be easily affected by the subdued light of that instrument.

In the year 1814, however, twelve years subsequent to the publication of Wedgwood's paper, M. Niépce, of Châlons, having directed his attention to the subject, succeeded in perfecting a process in which the Camera could be employed, although the sensibility was still so low that an exposure of some hours was required to produce the effect.

In the process of M. Niépce, which was termed "Heliography," or "sun-drawing," the use of the Silver Salts was discarded, and a resinous substance, known as "Bitumen of Judæa," substituted. This resin was smeared on the surface of a metal plate, and exposed to the luminous image. The light in acting upon it so changed its properties, that it became *insoluble* in certain essential oils. Hence, on subsequent treatment with the oleaginous sol-

vent, the *shadows* dissolved away, and the *lights* were represented by the unaltered resin remaining on the plate.

The Discoveries of M. Daguerre.—MM. Niépce and
Daguerre appear at one time to have been associated as
partners, for the purpose of mutually prosecuting their researches ; but it was not until after the death of the former, viz. in 1839, that the process named the Daguerreotype was given to the world. Daguerre was dissatisfied
with the slowness of action of the Bitumen sensitive surface, and directed his attention mainly to the use of the
Salts of Silver, which are thus again brought before our
notice.

Even the earlier specimens of the Daguerreotype, although far inferior to those subsequently produced, possessed a beauty which had not been attained by any Photographs prior to that time.

The sensitive plates of Daguerre were prepared by exposing a silvered tablet to the action of the vapour of
Iodine, so as to form a layer of Iodide of Silver upon the
surface. By a short exposure in the Camera an effect was
produced, not visible to the eye, but appearing when the
plate was subjected to the vapour of Mercury. This feature, viz. the production of a *latent* image upon Iodide of
Silver, with its subsequent development by a chemical reagent, is one of the first importance. Its discovery at once
reduced the time of taking a picture from hours to minutes,
and promoted the utility of the Art.

Daguerre also succeeded in *fixing* his proofs, by removal
of the unaltered Iodide of Silver from the shadows. The
processes employed however were imperfect, and the matter was not set at rest until the publication of a paper by
Sir John Herschel, on the property possessed by " Hyposulphites" of dissolving the Salts of Silver insoluble in
water.

*On a means of Multiplying Photographic Impressions,
and other Discoveries of Mr. Fox Talbot.*—The first communication made to the Royal Society by Mr. Fox Talbot,

in January, 1839, included only the preparation of a sensitive paper for copying objects by application. It was directed that the paper should be dipped first in solution of Chloride of Sodium, and then in Nitrate of Silver. In this way a white substance termed *Chloride of Silver* is formed, more sensitive to light than the Nitrate of Silver originally employed by Wedgwood and Davy. The object is laid in contact with the prepared paper, and, being exposed to light, a copy is obtained, which is *Negative,—id est*, with the light and shade reversed. A second sheet of paper is then prepared, and the first, or Negative impression, laid upon it, so as to allow the sun's light to pass through the transparent parts. Under these circumstances, when the Negative is raised, a natural representation of the object is found below; the tints having been again reversed by the second operation.

This production of a Negative Photograph, from which any number of Positive copies may be obtained, is a cardinal point in Mr. Talbot's invention, and one of great importance.

The patent issued for the process named *Talbotype* or *Calotype* dates from February, 1841. A sheet of paper is first coated with Iodide of Silver by soaking it alternately in Iodide of Potassium and Nitrate of Silver; it is then washed with solution of Gallic Acid containing Nitrate of Silver (sometimes termed *Gallo-Nitrate of Silver*), by which the sensibility to light is greatly augmented. An exposure in the Camera of some seconds or minutes, according to the brightness of the light, impresses an invisible image, which is brought out by treating the plate with a fresh portion of the mixture of Gallic Acid and Nitrate of Silver employed in exciting.

On the use of Glass Plates to retain Sensitive Films.— The principal defects in the Calotype process are attributable to the coarse and irregular structure of the fibre of paper, even when manufactured with the greatest care, and expressly for Photographic purposes. In consequence of this, the same amount of exquisite definition and sharpness of outline as that resulting from the use of metal plates, cannot be obtained.

We are indebted to Sir John Herschel for the first employment of glass plates to receive sensitive Photographic films.

The Iodide of Silver may be retained upon the glass by means of a layer of Albumen or white of egg, as proposed by M. Niépce de Saint-Victor, nephew to the original discoverer of the same name.

A more important improvement still is the employment of "Collodion" for a similar purpose.

Collodion is an ethereal solution of a substance almost identical with Gun-cotton. On evaporation it leaves a transparent layer, resembling goldbeater's skin, which adheres to the glass with some tenacity. M. Le Grey of Paris originally suggested that this substance might possibly be rendered available in Photography, but our own countryman, Mr. Archer, was the first to carry out the idea practically. In a communication to 'The Chemist' in the autumn of 1851, this gentleman gave a description of the Collodion process much as it now stands; at the same time proposing the substitution of *Pyro*-gallic acid for the Gallic acid previously employed in developing the image.

At that period no idea could have been entertained of the stimulus which this discovery would render to the progress of the Art; but experience has now abundantly demonstrated, that, as far as all qualities most desirable in a Photographic process are concerned, none at present known can excel, or perhaps equal, the Collodion process.

CHAPTER II.

THE SALTS OF SILVER EMPLOYED IN PHOTOGRAPHY.

By the term *Salt* of Silver we understand that the compound in question contains Silver, but not in its elementary form; the metal is in fact in a state of chemical union with other elements which disguise its physical properties, so that the Salt possesses none of the external characters of the Silver from which it was produced.

Silver is not the only metal which forms Salts; there are Salts of Lead, Copper, Iron, etc. *Sugar of Lead* is a familiar instance of a Salt of Lead. It is a white crystalline body, easily soluble in water, the solution possessing an intensely sweet taste; chemical tests prove that it contains Lead, although no suspicion of such a fact could be entertained from a consideration of its general properties.

Common Salt, or Chloride of Sodium, which is the type of the salts generally, is constituted in a similar manner; that is to say, it contains a metallic substance, the characters of which are masked, and lie hid in the compound.

The contents of this Chapter may be arranged in three Sections: the first describing the Chemistry of the Salts of Silver; the second, the action of Light upon them; the third, the preparation of a sensitive surface, with experiments illustrating the formation of the Photographic image.

SECTION I.

Chemistry of the Salts of Silver.

The principal Salts of Silver employed in the Photographic processes are four in number, viz. Nitrate of Silver, Chloride of Silver, Iodide of Silver, and Bromide of Silver. In addition to these, it will be necessary to describe the Oxides of Silver.

THE PREPARATION AND PROPERTIES OF THE NITRATE OF SILVER.

Nitrate of Silver is prepared by dissolving metallic Silver in Nitric Acid. Nitric Acid is a powerfully acid and corrosive substance, containing two elementary bodies united in definite proportions. These are Nitrogen and Oxygen; the latter being present in greatest quantity.

Nitric Acid is a powerful solvent for the metallic bodies generally. To illustrate its action in that particular, as contrasted with other acids, place pieces of silver-foil in two test-tubes, the one containing dilute Sulphuric, the other dilute Nitric Acid; on the application of heat a violent action soon commences in the latter, but the former is unaffected. In order to understand this, it must be borne in mind that when a metallic substance dissolves in an acid, the nature of the solution is different from that of an aqueous solution of salt or sugar. If salt water be boiled down until the whole of the water has evaporated, the salt is recovered with properties the same as at first; but if a similar experiment be made with a solution of Silver in Nitric Acid, the result is different: in that case *metallic* Silver is not obtained on evaporation, but Silver *combined with Oxygen and Nitric Acid*, both of which are strongly retained, being in fact in a state of chemical combination with the metal.

If we closely examine the effects produced by treating Silver with Nitric Acid, we find them to be of the following nature:—first, a certain amount of Oxygen is imparted to the metal, so as to form an *Oxide*, which Oxide dissolves in another portion of the Nitric Acid, producing *Nitrate* of the Oxide, or, as it is shortly termed, Nitrate of Silver.*

It is the *instability* of Nitric Acid therefore—its proneness to part with Oxygen—which renders it superior to the Sulphuric and to most acids in dissolving Silver and various other substances, both organic and inorganic.

Properties of Nitrate of Silver.—In preparing Nitrate of Silver, when the metal has dissolved, the solution is boiled down and set aside to crystallize. The salt however as so obtained is still acid to test-paper, and requires either re-crystallization, or careful heating to about 300° Fahrenheit. It is this retention of small quantities of Nitric Acid, and sometimes probably of Nitrous Acid, which renders much of the commercial Nitrate of Silver useless for Photography, until rendered neutral by fusion and a second crystallization.

Pure Nitrate of Silver occurs in the form of white crystalline plates, which are very heavy and dissolve readily in an equal weight of cold water. The solubility is much lessened by the presence of free Nitric Acid, and in the *concentrated* Nitric Acid the crystals are almost insoluble. Boiling Alcohol takes up about one-fourth part of its weight of the crystallized Nitrate, but deposits nearly the whole on cooling. Nitrate of Silver has an intensely bitter and nauseous taste; acting as a caustic, and corroding the skin by a prolonged application. Its aqueous solution does not redden blue litmus-paper.

Heated in a crucible the salt melts, and when poured into a mould and solidified, forms the white *lunar caustic* of commerce. At a still higher temperature it is decomposed, and bubbles of Oxygen Gas are evolved: the melted mass cooled and dissolved in water leaving behind a black powder, and yielding a solution, which is faintly alkaline

* The preparation of Nitrate of Silver from the standard coin of the realm is described in Part III., Art. " Silver."

to test-paper, from the presence of minute quantities of Nitrite or basic Nitrite of Silver.*

THE CHEMISTRY OF THE CHLORIDES OF SILVER.

Preparation of Protochloride of Silver.—The ordinary white Chloride of Silver may be prepared in two ways,— by the direct action of Chlorine upon metallic Silver, and by double decomposition between two salts.

If a plate of polished silver be exposed to a current of Chlorine Gas,† it becomes after a short time coated on the surface with a superficial film of white powder. This powder is Chloride of Silver, containing the two elements Chlorine and Silver united in single equivalents.

Preparation of Chloride of Silver by double decomposition.—In order to illustrate this, take a solution in water of Chloride of Sodium or "common salt," and mix it with a solution containing Nitrate of Silver; immediately a dense, curdy, white precipitate falls, which is the substance in question.

In this reaction the elements change places; the Chlorine leaves the Sodium with which it was previously combined, and crosses over to the Silver; the Oxygen and Nitric Acid are released from the Silver, and unite with the Sodium; thus

Chloride of Sodium *plus* Nitrate of Silver
equals Chloride of Silver *plus* Nitrate of Soda.

This interchange of elements is termed by chemists *double decomposition;* further illustrations of it, with the conditions necessary to the proper establishment of the process, are given in the first Chapter of Part III.

The essential requirements in two salts intended for the

---

* Nitrite of Silver differs from the Nitrate in containing less Oxygen, and is formed from it by the abstraction of two atoms of that element; it is described in the vocabulary, Part III.

† For the properties of the element "Chlorine," see the third division of the Work.

preparation of Chloride of Silver, are simply that the first should contain Chlorine, the second Silver, and that both should be soluble in water; hence the Chloride of Potassium or Ammonium may be substituted for the Chloride of Sodium, and the Sulphate or Acetate for the Nitrate of Silver.

In preparing Chloride of Silver by double decomposition, the white clotty masses which first form must be washed repeatedly with water, in order to free them from soluble Nitrate of Soda, the other product of the change. When this is done, the salt is in a pure state, and may be dried, etc., in the usual way.

Properties of Chloride of Silver.—Chloride of Silver differs in appearance from the Nitrate of Silver. It is not usually crystalline, but forms a soft white powder resembling common chalk or whiting. It is tasteless and insoluble in water; unaffected by boiling with the strongest Nitric Acid, but sparingly dissolved by concentrated Hydrochloric Acid.

Ammonia dissolves Chloride of Silver freely, as do solutions of Hyposulphite of Soda and Cyanide of Potassium. Concentrated solutions of alkaline Chlorides, Iodides, and Bromides are likewise solvents of Chloride of Silver, but to a limited extent, as will be more fully shown in Chapter IV., when treating of the modes of fixing the Photographic proofs.

Dry Chloride of Silver carefully heated to redness fuses, and concretes on cooling into a tough and semitransparent substance, which has been termed *horn silver* or *luna cornea*.

Placed in contact with metallic Zinc or Iron acidified with dilute Sulphuric Acid, Chloride of Silver is reduced to the metallic state, the Chlorine passing to the other metal under the decomposing influence of the galvanic current which is established.

Preparation and Properties of the Subchloride of Silver. —If a plate of polished Silver be dipped in solution of Perchloride of Iron, or of Bichloride of Mercury, a *black stain* is produced, the Iron or Mercury Salt losing a portion of Chlorine, which passes to the Silver and converts it super- ficially into Subchloride of Silver. This compound differs from the white Chloride of Silver in containing less Chlo- rine; the composition of the latter being represented by the formula $AgCl$, that of the former may perhaps be written as Ag_2Cl (?).

Subchloride of Silver is interesting to the Photographer as corresponding in properties and composition with the ordinary Chloride of Silver blackened by light. It is a pulverulent substance of a bluish-black colour not easily affected by Nitric Acid but decomposed by fixing agents such as Ammonia, Hyposulphite of Soda, or Cyanide of Potassium, into Chloride of Silver which dissolves, and in- soluble metallic Silver.

THE CHEMISTRY OF IODIDE OF SILVER.

The properties of *Iodine* are described in the third divi- sion of the Work: they are analogous to those of Chlorine and Bromine, the Silver Salts formed by these elements bearing also a strong resemblance to each other.

Preparation and Properties of Iodide of Silver.—Iodide of Silver may be formed in an analogous manner to the Chloride, viz. by the direct action of the vapour of Iodine upon metallic Silver, or by double decomposition, between solutions of Iodide of Potassium and Nitrate of Silver.

When prepared by the latter mode it forms an impal- pable powder, the colour of which varies slightly with the manner of precipitation. If the Iodide of Potassium be in excess, the Iodide of Silver falls to the bottom of the vessel nearly white; but with an excess of Nitrate of Silver it is of a straw-yellow tint. This point may be noticed, because the yellow salt is the one adapted for Photographic use, the other being insensible to the influence of light.

Iodide of Silver is tasteless and inodorous; insoluble in

water and in dilute Nitric Acid. It is scarcely dissolved by Ammonia, which serves to distinguish it from the Chlo-ride of Silver, freely soluble in that liquid. Hyposulphite of Soda and Cyanide of Potassium both dissolve Iodide, of Silver; it is also soluble in solutions of the alkaline Bromides and Iodides, as will be further explained in Chapter IV.

Iodide of Silver is reduced by Metallic Zinc in the same manner as the Chloride of Silver, forming soluble Iodide of Zinc and leaving a black powder.

THE PREPARATION AND PROPERTIES OF BROMIDE OF SILVER.

This substance so closely resembles the corresponding salts containing Chlorine and Iodine, that a short notice of it will suffice.

Bromide of Silver is prepared by exposing a silvered plate to the vapour of Bromine, or by adding solution of Bromide of Potassium to Nitrate of Silver. It is an insoluble substance, slightly yellow in colour, and distinguished from Iodide of Silver by dissolving in strong Ammonia and in Chloride of Ammonium. It is freely soluble in Hyposulphite of Soda and in Cyanide of Potassium.

The properties of the element Bromine are described in Part III.

CHEMISTRY OF THE OXIDES OF SILVER.

The Protoxide of Silver (Ag O).—If a little Potash or Ammonia be added to solution of Nitrate of Silver, an olive-brown substance is formed, which, on standing, collects at the bottom of the vessel. This is Oxide of Silver, displaced from its previous state of combination with Nitric Acid by the stronger oxide, Potash. Oxide of Silver is soluble *to a very minute extent* in pure water, the solution possessing an alkaline reaction to Litmus; it is easily dissolved by Nitric or Acetic Acid, forming a neutral Nitrate

c

or Acetate; also soluble in Ammonia (Ammonio-Nitrate of Silver), and in Nitrate of Ammonia, Hyposulphite of Soda, and Cyanide of Potassium. Long exposure to light converts it into a black substance, which is probably a Suboxide.

The Suboxide of Silver (Ag_2O ?)—This substance was obtained by Faraday on exposing a solution of the Ammonio-Nitrate of Silver to the action of the air. It bears a relation to the ordinary brown Protoxide of Silver similar to that which the Subchloride bears to Protochloride of Silver.

Suboxide of Silver is a black or grey powder, which assumes the metallic lustre on rubbing, and when treated with dilute Acids is resolved into Protoxide of Silver which dissolves, and metallic Silver.

SECTION II.

On the Photographic Properties of the Salts of Silver.

In addition to the Salts of Silver described in the first Section of this Chapter there are many others well known to chemists, as the Acetate of Silver, the Sulphate, the Citrate of Silver, etc. Some occur in crystals which are soluble in water, whilst others are pulverulent and insoluble.

The Salts of Silver formed by colourless Acids are white when first prepared, and remain so if kept in a dark place; but they possess the remarkable peculiarity of being darkened in colour by exposure to Light.

Action of Light upon the Nitrate of Silver.—The Nitrate of Silver is one of the most permanent of the Silver salts. It may be preserved unchanged in the crystalline form, or in solution in distilled water, for an indefinite length of time, even when constantly exposed to the diffused light of day. This is partly explained by the nature of the acid with which Oxide of Silver is associated in the Salt; *Nitric* Acid, possessing strong oxidizing properties, being opposed to the darkening influence of Light upon the Silver compounds.

Nitrate of Silver may, however, be rendered susceptible to the influence of Light, by adding to its solution *organic matter*, vegetable or animal. The phenomena produced in this case are well illustrated by dipping a pledget of cotton-wool, or a sheet of white paper, in solution of Nitrate of Silver, and exposing it to the direct rays of the sun; it slowly darkens, until it becomes nearly black. The stains upon the skin produced by handling Nitrate of Silver are caused in the same way, and are seen most evidently when the part has been exposed to light.

The varieties of organic matter which especially facilitate the blackening of Nitrate of Silver are such as tend *to absorb Oxygen;* hence pure vegetable fibre, free from Chlorides, such, for instance, as the Swedish filtering-paper, is not rendered very sensitive by being simply brushed with solution of the Nitrate, but a little grape sugar added soon determines the decomposition.

Decomposition of Chloride, Bromide, and Iodide of Silver by Light.—Pure moist Chloride of Silver* changes slowly from white to *violet* on exposure to light. Bromide of Silver becomes of a grey colour, but is less affected than the Chloride. Iodide of Silver (if free from excess of Nitrate of Silver) does not alter in appearance by exposure even to the sun's rays, but retains its yellow tint unchanged. Of these three compounds therefore *Chloride* of Silver is the most readily acted on by light, and papers prepared with this salt will become far darker on exposure than others coated with Bromide or Iodide of Silver.

There are certain conditions which accelerate the action of light upon the Chloride of Silver. These are, first, *an excess of Nitrate of Silver*, and second, *the presence of organic matter*. Pure Chloride of Silver would be useless

---

* The Chloride here spoken of is the compound prepared by adding a soluble Chloride to a solution of Nitrate of Silver: the product of the direct action of Chlorine upon metallic Silver is sometimes insensitive to light.

as a Photographic agent, but a Chloride with excess of Nitrate is very sensitive. Even Iodide of Silver, ordinarily
unaffected, is blackened by light when moistened with a
solution of the Nitrate of Silver.*

Organic matter combined with Chloride and Nitrate of
Silver gives a still higher degree of sensibility, and in this
way the Photographic papers are prepared.

The blackening of Chloride of Silver by Light explained.
—This may be studied by suspending pure Chloride of Silver in distilled water, and exposing it to the sun's rays for
several days.   When the process of darkening has proceeded to some extent, the supernatant liquid is found to
contain *free Chlorine*, or, in place of it, *Hydrochloric Acid*
(H Cl), the result of a subsequent action of the Chlorine
upon the water.

The luminous rays appear to loosen the affinity of the
elements Chlorine and Silver for each other ; hence a portion of Chlorine is separated, and the white Protochloride
is converted into the violet *Sub*chloride of Silver.   If an
atom of Nitrate of Silver be present, the liberated Chlorine
unites with it, displacing Nitric Acid, and forming again
Chloride of Silver, which is decomposed in its turn. The
excess of Nitrate of Silver thus exerts an accelerating influence upon the darkening of Chloride of Silver, by rendering the chain of chemical affinities more complete, and
preventing an accumulation of Chlorine in the liquid, which
would be a check to the continuance of the action.

Action of Light upon organic Salts of Silver.—On adding
diluted Albumen, or white of egg, to solution of Nitrate
of Silver, a flocculent deposit forms which is a compound
of the animal matter with Protoxide of Silver, and is known
as "Albuminate of Silver." This substance is at first
quite white, but on exposure to light it turns to a brick-
red colour. The change which takes place is one of *de-*

---

* The reader will understand that the Acetate, Sulphate, or any other
soluble Salt of Silver, might be substituted for the Nitrate in this experiment,

oxidation, the Protoxide of Silver losing a portion of its Oxygen, and a Suboxide of Silver, the product of the reduction, remaining in union with the oxidized Albumen. The red compound may therefore be loosely designated as an Albuminate of Suboxide of Silver.

Gelatine does not precipitate Nitrate of Silver in the same manner as Albumen: but if a sheet of transparent Gelatine be allowed to imbibe a solution of the Nitrate, it becomes of a clear ruby-red tint on exposure to light, and a true chemical compound of Gelatine, or a product of its oxidation, with a low Oxide of Silver, is produced.

Caseine, the animal principle of milk, is coagulated by Nitrate of Silver, and the red substance formed on exposing the curds to light may be viewed as analogous in composition to the corresponding compounds with Albumen and Gelatine.

Many other organic salts of Silver are darkened by light. The white Citrate of protoxide of Silver changes to a red substance, reacting with chemical tests in the same manner as Wöhler's Citrate of suboxide of Silver, which he obtained by reducing the ordinary Citrate in Hydrogen Gas. Glycyrrhizin, the Sugar of Liquorice, also forms a white compound with Oxide of Silver which becomes brown or red in the sun's rays.*

SIMPLE EXPERIMENTS ILLUSTRATING THE ACTION OF LIGHT UPON A SENSITIVE LAYER OF CHLORIDE OF SILVER ON PAPER.

In the performance of the most simple experiments on the decomposition of Silver Salts by Light, the student may employ ordinary *test-tubes*, in which small quantities of the two liquids required for the double decomposition may be mixed together.

When however *concentrated* solutions are used in this

---

* For further particulars on the action of light upon the Salts of Silver associated with organic matter, see the Author's paper on the composition of the photographic image, in the eighth Chapter.

way, the insoluble Silver Salt falls in dense and clotted masses, which, exposed to the sun's rays, quickly blacken on the exterior, but the inside is protected, and remains white. It is of importance therefore in Photography that the sensitive material should exist in the form of *a surface*, in order that the various particles of which it is composed may each one individually be brought into relation with the disturbing force.

Full directions for the preparation of sensitive Photographic paper are given in the second division of this work. The following is the theory of the process:—A sheet of paper is treated with solution of Chloride of Sodium or Ammonium, and subsequently with Nitrate of Silver; hence results a formation of Chloride of Silver in a fine state of division, with an excess of Nitrate of Silver, the Silver bath having been purposely made stronger in proportion than the salting solution.

Illustrative Experiment No. I.—Place a square of sensitive paper (prepared according to the directions given in the Second Part of the work) in the direct rays of the sun, and observe the gradual process of darkening which takes place; the surface passes through a variety of changes in colour until it becomes of a deep chocolate-brown. If the Light is tolerably intense, the brown shades are probably reached in from three to five minutes; but the sensibility of the paper, and also the nature of the tints, will vary much with the character of the organic matter present.

Experiment No. II.—Lay a device cut from black paper upon a sheet of sensitive paper, and compress the two together by means of a sheet of glass. After a proper length of exposure the figure will be exactly copied, the tint however being reversed: the black paper protecting the sensitive Chloride beneath, produces a *white* figure upon a dark ground.

Experiment No. III.—Repeat the last experiment, substituting a piece of lace or gauze-wire for the paper device.

This is intended to show the minuteness with which objects can be copied, since the smallest filament will be distinctly represented.

Experiment No. IV.—Take an engraving in which the contrast of light and shade is tolerably well marked, and having laid it closely in contact with the sensitive paper, expose as before. This experiment shows that the surface darkens in degrees proportionate to the intensity of the light, so that the *half* shadows of the engraving are accurately maintained, and a pleasing gradation of tone produced.

In the darkening of Photographic papers, the action of the light is quite superficial, and although the black colour may be intense, yet the amount of reduced Silver which forms it is so small that it cannot conveniently be estimated by chemical reagents. This is well shown by the results of an analysis performed by the Author, in which the total weight of Silver obtained from a blackened sheet measuring nearly 24 by 18 inches amounted to less than *half a grain.* It becomes therefore of great importance in preparing sensitive paper to attend to the condition of the surface layer of particles, the action rarely extending to those beneath. The use of Albumen, Gelatine, etc., which will be explained in the eighth Chapter, has reference to this amongst other advantages, and secures a better and more sharply defined print.

CHAPTER III.

ON THE DEVELOPMENT OF AN INVISIBLE IMAGE BY MEANS OF A REDUCING AGENT.

It has been shown in the previous Chapter that the majority of the Salts of Silver, both organic and inorganic, are darkened in colour on exposure to light, and, by the loss of Oxygen, Chlorine, etc., become reduced to the conditions of *Sub*salts.

Many of the same compounds are also susceptible of a change under the influence of light, which is even more remarkable. This change takes place after a comparatively short exposure, and as it does not affect the appearance of the sensitive layer, for some time it escaped notice: but it was afterwards discovered that an impression, before invisible, might be brought out by treating the plate with certain chemical agents which are without effect on the original unchanged salt, but quickly *blacken* it after exposure.

It is a remarkable fact that the Silver compounds most readily affected by light alone, are not the most sensitive to the reception of the invisible image. Thus, of Photographic papers prepared with Chloride, Bromide, or Iodide of Silver, the former assume the deepest shade of colour under the influence of the sun's rays, but if all be exposed *momentarily*, and then removed, the greatest amount of effect will be developed upon the Iodide paper. Iodide of Silver therefore is the salt commonly used when sensibility is an object, but it should be noted that images nearly or quite latent can be impressed upon many other of the compounds of Silver, including those belonging to the animal and vegetable kingdoms.

Experiments illustrating the Formation of an Invisible Image.—Take a sheet of sensitive paper, prepared with Iodide of Silver by the method given in the fourth Chapter of Part II., and having divided it into two parts, expose one of them to the luminous rays for a few seconds. No visible decomposition takes place, but on removing the pieces to a room dimly illuminated, and brushing with a solution of *Gallic Acid*, a manifest difference will be observed; the one being unaffected, whilst the other darkens gradually until it becomes black.

Experiment II.—A prepared sheet is shielded in certain parts by an opaque substance, and then after the requisite exposure, which is easily ascertained by a few trials, treated with the Gallic Acid as before; in this case the protected part remains white, whilst the other darkens to a greater or less extent.

In the same way, copies of leaves, engravings, etc. may be made, very correct in the shading and much resembling those produced by the prolonged action of light alone upon the Chloride of Silver.

The object of employing a substance like Gallic Acid to *develope* or bring out to view an invisible image, in preference to forming the picture by the direct action of light, unassisted by a developer, is the *economy of time* thereby effected. This is well shown in the results of some experiments conducted by M. Claudet in the Daguerreotype process: he found that with a sensitive layer of Bromo-Iodide of Silver, an intensity of light three thousand times greater was required if the use of a developer was omitted, and the exposure continued until the picture became visible upon the plate.

To increase the sensitiveness of Photographic preparations is a point of great consequence; and indeed, when the Camera is used, from the low intensity of the luminous image formed in that instrument, no other plan than the one above described would be practicable. Hence the advancement, and indeed the very origin, of the Photographic Art, may be dated from the first discovery of a process for bringing out to view an invisible image by means of a reducing agent.

The present Chapter is divided into three Sections: —first, the chemical properties of the substances usually employed as developers;—second, their mode of action in reducing the Salts of Silver;—third, hypotheses on the action of light in impressing a latent image.

SECTION I.

Chemistry of the various Substances employed as Developers.

Development is essentially a process of *reduction*, or, in other words, of *deoxidation*. If we take a certain metal, we can, by means of Nitric Acid, impart Oxygen to it, so that it becomes first an Oxide, and afterwards, by solution of the Oxide in the excess of acid, *a salt*. When this salt is formed, by a series of chemical operations the reverse of the former it may be deprived of all its Oxygen, and the metallic element again isolated.

The degree of facility with which oxidation as well as reduction is performed, depends upon the affinity for Oxygen which the particular metal under treatment possesses. In this respect there is considerable difference, as may be shown by a reference to the two well-known metals, Iron and Gold. How speedily does the first become tarnished and covered with rust, whilst the other remains bright even in the fire! It is indeed possible, by a careful process, to form Oxide of Gold; but it retains its Oxygen so loosely that the mere application of heat is sufficient to drive it off, and leave the metal in a pure state.

Silver, Gold, and Platinum all belong to the class of *noble* metals, having the least affinity for Oxygen: hence their Oxides are unstable, and any body tending strongly to absorb Oxygen will reduce them to the metallic state.

Observe, therefore, that the substances employed by the Photographer to assist the action of the light, and to develope the picture, act by removing Oxygen. The sensitive Salt of Silver is thus *reduced*, more or less completely, in the parts touched by light, and an opaque deposit results which forms the image.*

The most important of the developers are as follows:— Gallic Acid, Pyrogallic Acid, and the *Proto*salts of Iron.

CHEMISTRY OF GALLIC AND PYROGALLIC ACIDS.

a. *Of Gallic Acid.*—Gallic Acid is obtained from *Gall Nuts*, which are peculiar excrescences formed upon the branches and shoots of the *Quercus infectoria* by the puncture of a species of insect. The best kind is imported from Turkey, and sold in commerce as Aleppo Galls. Gall Nuts do not contain Gallic Acid ready formed, but an analogous chemical principle termed *Tannic Acid*, well known for its astringent properties and employment in the process of tanning raw hides.

Gallic Acid is produced by the *decomposition and oxidation* of Tannic Acid when powdered galls are exposed for a long time in a moist state to the action of the air. By boiling the mass with water and filtering whilst hot, the acid is extracted, and crystallizes on cooling, on account of its sparing solubility in cold water.

Gallic Acid occurs in the form of long silky needles, soluble in 100 parts of cold and 3 of boiling water; they are also readily soluble in Alcohol, but sparingly in Ether.

* These remarks do not apply to the vapour of Mercury employed as a developing agent in the Daguerreotype. The chemistry of that process will be explained in a separate Chapter.

The aqueous solution becomes mouldy on keeping, to obviate which the addition of Acetic Acid or a drop or two of Oil of Cloves is recommended.

Gallic Acid is a feeble acid, scarcely reddening litmus; it forms salts with the alkaline and earthy bases, such as Potash, Lime, etc., but not with the oxides of the noble metals. When added to Oxide of Silver the metallic element is separated and the Oxygen absorbed.

b. *Pyrogallic Acid.*—The term *pyro* prefixed to Gallic Acid implies that the new substance is obtained by the *action of heat* upon that body. At a temperature of about 410° Fahr., Gallic Acid is decomposed, and a white sublimate forms, which condenses in lamellar crystals; this is Pyrogallic Acid.

Pyrogallic Acid is very soluble in cold water, and in Alcohol and Ether; the solution decomposes and becomes brown by exposure to the air. It gives an indigo blue colour with Protosulphate of Iron, which changes to dark green if any Persulphate be present.

Although termed an *acid*, this substance is strictly *neutral;* it does not redden litmus-paper, and forms no salts. The addition of Potash or Soda decomposes Pyrogallic Acid, at the same time increasing the attraction for Oxygen; hence this mixture may conveniently be employed for absorbing the Oxygen contained in atmospheric air. The compounds of Silver and Gold are reduced by Pyrogallic Acid even more rapidly than by Gallic Acid, the reducing agent absorbing the Oxygen, and becoming converted into Carbonic Acid and a brown matter insoluble in water.

Commercial Pyrogallic Acid is often contaminated with empyreumatic oil, and also with a black insoluble substance known as *Metagallic Acid,* which is formed when the heat is raised above the proper temperature in the process of manufacture.

CHEMISTRY OF THE PROTOSALTS OF IRON.

The combinations of Iron with Oxygen are somewhat numerous. There are two distinct Oxides which form Salts, viz. the Protoxide of Iron, containing an atom of Oxygen to one of metal; and the Peroxide, with an atom and a half of Oxygen to one of metal. As *half atoms* however are not allowed in chemical language, it is usual to say that the Peroxide of Iron contains three equivalents of Oxygen to two of metallic Iron.

Expressed in symbols, the composition is as follows:—

Protoxide of Iron, Fe O.

Peroxide of Iron, Fe_2O_3.

The Proto- and Persalts of Iron do not resemble each other in their physical and chemical properties. The former are usually of an apple-green colour, and the aqueous solutions almost colourless, if not highly concentrated. The latter, on the other hand, are dark, and give a yellow or even blood-red solution.

The Protosalts of Iron are alone useful in Photography; but the following experiment will serve to illustrate the properties of both classes of salts:—Take a crystal of Protosulphate of Iron, and, having reduced it to powder, pour a little Nitric Acid upon it in a test-tube. On the application of heat, abundance of fumes will be given off, and a red solution obtained. The Nitric Acid in this reaction imparts Oxygen, and converts the *Proto*sulphate entirely into a *Per*sulphate of Iron. It is this feature, viz. the tendency to absorb Oxygen, and to pass into the state of Persalts, which makes the Protosalts of Iron useful as developers.

There are two Protosalts of Iron commonly employed by Photographers: the Protosulphate and the Protonitrate of Iron.

a. *Protosulphate of Iron.*—This salt, often termed *Copperas* or *Green Vitriol*, is an abundant substance, and used for a variety of purposes in the arts. Commercial

Sulphate of Iron however, being prepared on a large scale, requires recrystallization to render it sufficiently pure for Photographic purposes.

Pure Sulphate of Iron occurs in the form of large transparent, prismatic crystals, of a delicate green colour: by exposure to the air they gradually absorb Oxygen and become rusty on the surface. Solution of Sulphate of Iron, colourless at first, afterwards changes to a red tint, and deposits a brown powder; this powder is a *basic* Persulphate of Iron, that is, a Persulphate containing an excess of the oxide or *base*. By the addition of Sulphuric or Acetic Acid to the solution, the formation of a deposit is prevented, the brown powder being soluble in acid liquids.

The Crystals of Sulphate of Iron include a large quantity of water of crystallization, a part of which they lose by exposure to dry air. By a higher temperature, the salt may be rendered perfectly *anhydrous,* in which state it forms a white powder.

b. *Protonitrate of Iron.*—This salt is prepared by double decomposition between Nitrate of Baryta or of Lead and Protosulphate of Iron. It is an unstable substance and crystallizes with great difficulty; its aqueous solution is pale green at first, but very prone to decomposition, even more so than the corresponding Sulphate of Iron.

SECTION II.

The Reduction of Salts of Silver by Developing Agents.

The general theory of the reduction of metallic oxides having been explained, it may be desirable to enter more minutely into the exact nature of the process as applied to the compounds of Silver.

First, the Reduction of the Oxide of Silver will be taken, as the most simple illustration; then that of Salts of Silver formed by Oxygen-acids; and lastly, of the Chloride, Iodide, and Bromide of Silver containing no Oxygen.

Reduction of Oxide of Silver.—To illustrate this conveniently, the Oxide of Silver should be in a state of solution; water dissolves Oxide of Silver very sparingly, but it is freely soluble in Ammonia, forming the liquid known as Ammonio-Nitrate of Silver. If, therefore, a little of the Ammonio-Nitrate of Silver be placed in a test-tube, and solution of Sulphate of Iron be added to it, immediately it becomes discoloured, and a deposit settles to the bottom.

This deposit is metallic Silver, produced by the reducing agent appropriating to itself the Oxygen previously combined with the metal. As metallic Silver does not dissolve in Ammonia, the liquid becomes turbid, and the metal subsides in the form of a bulky precipitate.

Reduction of the Oxyacid Salts of Silver.—The term *Oxyacid* includes those salts which contain the Oxide of Silver intimately combined with Oxygen-acids ; as *e. g.* the Nitrate of Silver, the Sulphate, the Acetate of Silver, etc.

These salts, soluble in water, are reduced by developing agents in the same manner as Oxide of Silver, but more slowly. The presence of an acid united with the base is a hindrance to the process and tends to keep the oxide in solution, especially when that acid is powerful in its affinities. To illustrate the effect of the acid constituent of the salt in retarding reduction, take two test-tubes, the one containing Ammonio-Nitrate, and the other ordinary Nitrate of Silver—a single drop of solution of Sulphate of Iron added to each will indicate an evident difference in the rapidity of deposition.

The precipitate of metallic Silver obtained by the action of reducing agents upon the Nitrate, varies much in colour and in general appearance. If Gallic or Pyrogallic Acid be employed, it is a black powder ;* whilst the salts of Iron, and especially the same with free Nitric Acid add-

---

* Silver precipitated by Gallic or Pyrogallic Acid does not appear to be free from organic matter, and probably contains also a small proportion of Oxygen.

ed, produce a sparkling precipitate, resembling what is termed *frosted silver.* Grape Sugar and many of the essential oils, such as the Oil of Cloves, etc., separate the metal from Ammonio-Nitrate of Silver in the form of a brilliant mirror film, and are often employed in silvering glass.

In remarking upon these peculiarities in the molecular condition of precipitated Silver, it should be observed that the appearance of a metal whilst in mass is no indication of its colour when in the state of fine powder. Platinum and Iron, both bright metals, and susceptible of a high polish, are dull and intensely black when in a fine state of division; Gold is of a purple or yellowish brown; Mercury a dirty grey.

Reduction of the Hydracid Salts of Silver.—By the term *Hydracid* is meant Salts of Silver which contain no Oxygen or Oxygen-acids, but simply elements like Chlorine or Iodine combined with Silver. These elements are characterized by forming acids with Hydrogen, which acids are hence called *Hydr*acids. Hydrochloric Acid (HCl) is an example; so also is Hydriodic Acid (HI).

The reduction of the Hydracid Salts requires to be discussed separately, because it is evidently different from that already described; the reducing agent tending only to absorb *Oxygen*, which is not present in these salts. The explanation is as follows: When a Chloride of a noble metal is reduced by a developer, *an atom of water*, composed of Oxygen and Hydrogen, takes a part in the reaction. The Oxygen of the water passes to the developer, the Hydrogen to the Chlorine.

To illustrate this, take a solution of Chloride of Gold, and add to it a little Sulphate of Iron. A yellow deposit of metallic Gold soon forms, and the supernatant liquid is found, by testing, to be acid from free Hydrochloric Acid. The following simple diagram, in which however the *number* of the atoms concerned is omitted, may assist the comprehension of the change.

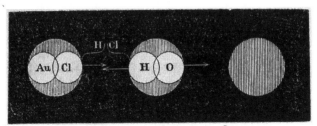

Compound Atom of   Compound Atom   Atom of
Chloride of Gold.   of Water.   Sulphate of Iron.

The symbol Au represents Gold, Cl Chlorine, H Hydrogen, and O Oxygen. Observe that the molecules H and O separate from each other and pass in opposite directions: the latter unites with the Sulphate of Iron; the former meets Cl, and produces Hydrochloric Acid (HCl), whilst the atom of Gold is left alone.

Hence there is no theoretical difficulty in supposing a reduction of Iodide of Silver by a developer, if we associate with the Iodide an atom of water to furnish the Oxygen. Unless the sensitive plate however has been exposed to the light, the reduction does not readily take place; nor can it be produced under any circumstances, with or without light, when the whole of the *free Nitrate of Silver* has been washed away from the plate. *Pure* Iodide of Silver is therefore unaffected by a developer, and the compound which blackens on the application of Sulphate of Iron or Pyrogallic acid is an Iodide with excess of Nitrate of Silver.

The mode in which a Salt of Silver, such as the Nitrate,

Compound Atom of   Compound Atom of   Atom of
Iodide of Silver.   Nitrate of Silver.   Sulphate of Iron.

soluble in water, may act in facilitating the reduction of Iodide of Silver, is shown in the preceding diagram, which corresponds closely with the last.

Notice that the compound atom of Nitrate of Silver contains a molecule of Oxygen for the developer, one of Silver (Ag) for the separated Iodine, and an atom of Nitric Acid, (NO_5), which is liberated, and takes no further part in the change.

The chain of chemical affinities is more complete in this diagram than in the last, where an atom of water only was present, the affinity of Iodine for Silver being greater than that of Iodine for Hydrogen. Hence it is possible that an excess of Nitrate of Silver may, by furnishing an elementary basis for which Iodine has an attraction, assist in *drawing off* that element, so to speak, from the original particle of Iodide of Silver touched by light.*

SECTION III.

The formation and development of the Latent Image.

It was shown in the second Chapter that the continued action of white light upon certain of the Salts of Silver resulted in the separation of elements like Chlorine and Oxygen and the partial *reduction* of the compound. We have also seen that bodies possessing affinity for Oxygen, such as Sulphate of Iron and Pyrogallic Acid, tend to produce a similar effect; acting in some cases with great energy and precipitating metallic Silver in a pure state.

In forming an extemporaneous theory on the production

---

* The reader must not suppose from the remarks which have been made in this Section that images obtained by development consist invariably of pure metallic Silver. It can be shown that such is not the case,—that the process of reduction is in many cases suspended when a part only of the Oxygen has been removed; and hence results a *subsalt* similar to that produced by the direct action of light upon organic compounds of Silver, and differing in properties from metallic Silver. For further particulars see the Author's Photographic researches in the eighth Chapter.

of the latent image in the Camera, it would therefore be natural to suppose that the process consisted in setting up a reducing action upon the sensitive surface by means of light, afterwards to be continued by the application of the developing solution. This idea is to a certain extent correct, but it requires some explanation. The effects produced by the light and the developer are not so precisely similar that the one agency can always be substituted for the other: an insufficient exposure in the Camera cannot be remedied by prolonging the development of the image. In the Photographic processes on paper it is indeed found that a certain latitude may be allowed; but, as a rule, it should be stated that a definite time is occupied in the formation of the invisible image, which may not be shortened or extended beyond its proper limits with impunity. There is a maximum point beyond which no advance is made; hence if the plate be not then removed from the Camera, those portions of the image formed by the brightest lights are speedily overtaken by the "half tones," so that, on developing, an image appears without that contrast between lights and shadows which is essential to the artistic effect. On the other hand, in a case of insufficient exposure, the feeble rays of light not having been allowed time to impress the plate, the half shadows cannot be brought out on subsequent treatment with the developing agent.

A careful study of the phenomena involved in this part of the process cannot fail to show that the ray of Light determines a *molecular* change of some kind in the particles of Iodide of Silver forming the sensitive surface. This change is not of a nature to alter the composition or the chemical properties of the salt. The Iodine does not leave the surface, or there would be a difference in the appearance of the film, or in its solubility in Hyposulphite of Soda.

The following diagrams may perhaps be useful in mechanically illustrating what is meant by a molecular change.

Fig. 1 represents a compound molecule of Iodide of

Silver, the component atoms of which are closely associated.

Fig. 2. The same after the action of a disturbing force. The simple molecules have not altogether separated, but they are prepared to do so, touching only at a single point.

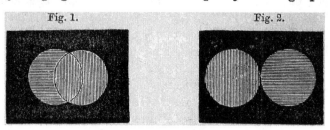

Fig. 1.   Fig. 2.

Now the effect produced on this combination by a developer is understood, if we suppose that in the first case the affinity of the Iodine for Silver is too great to allow of its separation; but in the second, this affinity having been loosened, the structure gives way, and metallic Silver is the result.

This hypothesis has the merit of simplicity, and is not opposed to known facts; it may therefore for the present be received. The point however on which a doubt must rest is—whether the molecular disturbance produced by light upon Iodide of Silver leads to a reduction of that Salt by the developer. No image can be produced on the application of Pyrogallic Acid *unless the particles of Iodide are in contact with Nitrate of Silver;* and hence it may be the Nitrate and not the Iodide which is reduced—that is, the impressed molecule of Iodide may determine the decomposition of a contiguous particle of Nitrate, itself remaining unchanged. This view is supported to some extent by Moser's experiments, shortly to be quoted; and also by the fact that the delicate image first formed can be *intensified* by treating it with a mixture of the developing solution and Nitrate of Silver, even after the Iodide has been removed by a fixing agent. The following experiment will serve to illustrate this.—

Take a sensitive Collodion plate, and having impressed an invisible image upon it by a proper exposure in the Camera, remove it to the dark room, and pour over it the solution of Pyrogallic Acid. When the picture has fully appeared, stop the action by washing the plate with water, and remove the unaltered Iodide of Silver by Cyanide of Potassium. An examination of the image at this stage will show that it is perfect in the details, but pale and translucent. The plate is then to be taken back again to the dark room and treated with fresh Pyrogallic Acid, *to which Nitrate of Silver has* been added; immediately the picture becomes much blacker, and continues to darken, even to complete opacity, if the supply of Nitrate be kept up.

Now in this experiment it is evident that the additional deposit upon the image is produced from the *Nitrate* of Silver, the whole of the Iodide having been previously removed. Observe also, *that it forms only upon the image, and not upon the transparent parts of the plate.* Even if the Iodide, untouched by light, be allowed to remain, the same rule holds good;—the Pyrogallic Acid and Nitrate of Silver react upon each other and produce a metallic deposit; this deposit however has no affinity for the unaltered Iodide upon the part of the plate corresponding to the shadows of the picture, but attaches itself in preference to the Iodide already blackened by light.

This second stage of the development, by which a feeble image may be strengthened and rendered more opaque, is sometimes termed "development by precipitation," and should be correctly understood by the practical operator.

Researches of M. Moser.—The papers of M. Ludwig Moser 'On the Formation and Development of Invisible Images,' published in 1842, explain so clearly many remarkable phenomena of occasional occurrence in the Collodion and paper processes, that no apology need be offered for referring to them somewhat at length.

His first proposition may be stated thus:—"If a polished surface has been touched in particular parts by any body, it acquires the property of precipitating certain vapours on these spots differently to what it does on the other untouched parts." To illustrate this, take a thin plate of metal, having characters *excised;* warm it gently, and lay it upon the surface of a clean mirror glass for a few minutes: then remove, allow to cool, and *breathe* upon the glass, when the outlines of the device will be distinctly seen. A plate of polished Silver may be substituted for the glass, and in place of developing the image by the breath, it may be brought out by Mercurial vapour.

The second proposition of M. Moser is as follows:—
"*Light* acts on bodies, and its influence may be tested by vapours that adhere to the substance."—A plate of mirror glass is exposed in the Camera to a bright and intense light; it is then removed and breathed upon, when an image before invisible will be developed, the breath settling most strongly upon the parts where the light has acted. A plate of polished Silver may be used as before instead of glass, the vapour of Mercury or of water being employed to develope the image. An *iodized Silver plate* is still more sensitive to the influence of the light, and receives a very sharp and perfect impression under the action of the Mercury.

It seems therefore from these experiments and others not quoted, that the surfaces of various bodies are capable of being modified by contact with each other, or by contact with a ray of light, in such a way as to impart an affinity for a vapour; and further, that many of the Salts of Silver are in the list of substances admitting of such modification. But it is also evident that the same condition of surface which causes a vapour to settle in a peculiar manner also affects the behaviour of the Silver Salt when treated with a reducing agent. Thus, if a clean glass plate be touched in certain spots by the warm finger, the impression soon disappears, but is again seen on breathing upon the glass; and if this same plate be coated with a very delicate layer of Iodized Collodion and passed through the Nitrate bath, the solution of Pyrogallic acid will commonly produce a well-defined outline of the figure even before the plate has been exposed to the light. This experiment, although it does not invariably succeed, is nevertheless an instructive one, and shows the necessity of cleaning the plates used in Photography with care. If there be any irregularity in the manner in which the breath settles upon the glass when it is breathed on, a condition of surface exists at that point which will probably so modify the layer of Iodide of Silver, that the action of the developing fluid will be in some way interfered with.

One more remarkable fact observed by M. Moser may be quoted. He finds that the action of light upon the Daguerreotype plate is of an *alternating* kind: it first gives an affinity for Mercury, and then removes it. "If light acts on Iodide of Silver," he says, "it imparts to it the power of condensing mercurial vapours; but if it acts beyond a certain time, it then diminishes this power and at length takes it away altogether." This is precisely in accordance with phenomena observed also in the Collodion process, where the deposit of metallic Silver is sometime less marked than usual if the plate has been exposed in the Camera beyond the proper period of time.

A curious perversion of the developing process is occasionally met with, in which on the application of the Pyrogallic Acid, the deposit of Silver takes place upon the *shadows* of the picture, and not upon the lights; hence on viewing the image by transmitted light, the usual appearance is reversed. This may perhaps be explained by an alternating action of the light as above suggested.

A phenomenon at first sight even more remarkable has occurred, in which, on developing the plate, *two* images start out instead of one. The secondary image in such a case is probably the remains of a previous impression which, although apparently removed by washing, had nevertheless modified the surface of the glass so as to affect the layer of Iodide of Silver; and if the glass were *breathed* upon before again coating it with Collodion, there is every reason to suppose that the outlines of the accidental image would be seen.*

* Since writing the above, the Author has perused with pleasure a paper by Mr. Grove on the production of latent images by electricity, with a mode of fixing them. In the experiments described, a plate of glass, electrized in certain portions only, was breathed upon, or exposed to the fumes of Hydrofluoric Acid. In either case the vapour settled exclusively upon the non-electrical part of the glass, thus developing a latent image. When the plate was first submitted to electrization, and then coated with Iodide of Silver upon Collodion, and exposed to light,—solution of Pyrogallic Acid produced a reduction of Silver *only* upon the parts of the glass corresponding to those on which the breath settled in the previous experiment; thus indicating that the electricity neutralized the effect of light upon the sensitive Iodide of Silver.

CHAPTER IV.

ON FIXING THE PHOTOGRAPHIC IMAGE.

A SENSITIVE layer of Chloride or Iodide of Silver on which an image has been formed, either with or without the aid of a developing agent, must pass through further treatment in order to render it indestructible by diffused light.

It is true that the image itself is sufficiently permanent, and cannot be said, in correct language, to need *fixing ;* but the unchanged Silver Salt which surrounds it, being still sensitive to light, tends to be decomposed in its turn, and so the picture is lost. It is therefore necessary to emove this salt by applying some chemical agent capable of dissolving it. The list of solvents of Chloride and Iodide of Silver has been given in Chapter II., but some are better adapted for fixing than others. In order that any body may be employed with success as a fixing agent, it is required not only that it should dissolve unchanged Chloride or Iodide of Silver, but that it should produce no injurious effect upon the same salts reduced by light.

This *solvent action upon the image,* as well as upon the parts which surround it, is most liable to happen when the agency of light alone, without a developer, has been employed. In that case the darkened surface, not being reduced perfectly to the metallic state, remains soluble to a certain extent in the fixing liquid.

CHEMISTRY OF THE VARIOUS FIXING AGENTS.

The following will be mentioned :—Ammonia—Alkaline Chlorides—Alkaline Iodides—Alkaline Hyposulphites—Alkaline Cyanides.

AMMONIA.

The·properties of the alkaline liquid "Ammonia" are given in Part III. Ammonia dissolves Chloride of Silver readily, but not Iodide of Silver: hence its use is necessarily confined to the paper proofs upon Chloride of Silver. Even these however cannot advantageously be fixed in. Ammonia unless a deposit of Gold has been previously produced upon the surface by a process of "toning," presently to be explained: a peculiar and unpleasant red tint is always caused by Ammonia acting upon the darkened material of a sun picture as it comes from the printing-frame: but this is obviated by the employment of the Gold.

ALKALINE CHLORIDES, IODIDES, AND BROMIDES.

The Chlorides of Potassium, Ammonium, and Sodium possess the property of dissolving a small portion of Chloride of Silver. In the act of solution a double salt is formed; that is, a compound of Chloride of Sodium with · Chloride of Silver, which may be crystallized out by allowing the liquid to evaporate spontaneously.

The earlier Photographers employed a saturated solution of common Salt for fixing paper prints; but the fixing action of the Alkaline Chlorides is slow and imperfect, and their use may now be said to be obsolete.

The Iodide and Bromide of Potassium have both been used as fixing agents. They dissolve Iodide of Silver, forming with it a double salt in the manner before described.

It is important to remark in the solution of the insoluble Silver Salts by Alkaline Chlorides, Iodides, etc., that the amount dissolved is not in proportion to the *quantity* of the solvent, but to the degree of concentration of its aqueous solution.   This is not usual with solvents which act by entering into chemical combination with the substance dissolved.   Commonly a given weight of the one salt dissolves a given weight of the other, independent of the amount of water present.   The peculiarity in the case before us depends upon the fact that the double salt formed is *decomposed* by a large quantity of water.   Hence it is a *saturated* solution of Chloride of Sodium which possesses the greatest power of fixing paper prints; and with the Bromide or Iodide of Potassium the same rule holds good—the stronger the solution the more Iodide of Silver will be taken up.   The addition of water produces milkiness and a deposit of the silver Salt previously dissolved.

ALKALINE HYPOSULPHITES.

Hyposulphurous Acid is one of the Oxides of Sulphur. It is, as its name implies, of an acid nature, and takes its place upon the list immediately below Sulphurous Acid ("*upo*," under).

The Hyposulphite of Soda commonly employed by Photographers is a neutral combination of Hyposulphurous Acid and the alkali Soda.   It is selected as being more economical in preparation than any other Hyposulphite adapted for fixing.

Hyposulphite of Soda occurs in the form of large translucent groups of crystals, which include five atoms of water.   These crystals are soluble in water almost to any extent, the solution being attended with the production of cold; they have a nauseous and bitter taste.

In the solution of Silver compounds by Hyposulphite of Soda *a double decomposition* always takes place; thus :—

Hyposulphite of Soda + Chloride of Silver
= Hyposulphite of Silver + Chloride of Sodium.

The Hyposulphite of Silver with an excess of Hyposulphite of Soda forms a soluble double salt, which may be crystallized out by evaporating the solution. It possesses an intensely sweet taste, and contains one atom of Hyposulphite of Silver, chemically combined with *two* of Hyposulphite of Soda. In addition to this there is a second double Salt, differing from the first in being *very sparingly* soluble in water. It is formed by acting upon Chloride of Silver with a solution of Hyposulphite of Soda already saturated, or nearly so, with Silver Salts; and contains single atoms of each constituent.

The fact that the Silver contained in an ordinary fixing Bath is present in the state of *Hyposulphite* must be borne in mind, because this salt is liable to undergo peculiar chemical changes, as will be better shown in Chapter VIII.

Iodide of Silver is dissolved by Hyposulphite of Soda more slowly than Chloride of Silver, and the amount eventually taken up is less. This is explained as follows :— During the solution of Iodide of Silver, *Iodide of Sodium* is formed, and this alkaline Iodide has a prejudicial effect upon the continuance of the process. *Chloride* of Sodium has not the same action, neither has Bromide of Sodium, consequently the corresponding Silver Salts dissolve to a greater extent than the Iodide.

ALKALINE CYANIDES.

The chemistry of Cyanogen is sketched in Part III.

The Cyanide of *Potassium* is the salt most frequently employed in fixing. It occurs in commerce in the form of fused lumps of considerable size. In this state it is usually contaminated with a large percentage of Carbonate of Potash, amounting in some cases to more than half its weight. By boiling in proof Spirit the Cyanide may be extracted and crystallized, but this operation is scarcely required as far as its use in Photography is concerned.

Cyanide of Potassium absorbs moisture on exposure to

the air. It is very soluble in water, but the solution decomposes on keeping; changing in colour and evolving the odour of *Prussic Acid*, which is a Cyanide of Hydrogen. Cyanide of Potassium is highly poisonous, and must be used with caution.

Solution of Cyanide of Potassium is a most energetic agent in dissolving the insoluble Silver Salts: far more so, in proportion to the quantity used, than the Hyposulphite of Soda. The Salts are in all cases converted into Cyanides, and exist in the solution in the form of soluble double Salts, which, unlike the double Iodides, are not affected by dilution with water. Cyanide of Potassium is unadapted for fixing positive proofs upon Chloride of Silver; and even when a developer has been used, unless the solution is tolerably dilute, it is apt to attack the image and dissolve it.

CHAPTER V.

ON THE NATURE AND PROPERTIES OF LIGHT.

THE present Chapter is devoted to a discussion of the more remarkable properties of Light; the object being to select certain prominent points, and to state them as clearly as possible, referring, for information of a more complete kind, to acknowledged works on the subject of Optics.

The Chapter will be divided into five Sections:—first, the compound nature of Light; second, the laws of refraction of Light; third, the construction of Lenses and of the Camera; fourth, the Photographic action of coloured Light; fifth, on Binocular Vision and the Stereoscope.

SECTION I.

The Compound Nature of Light.

The ideas entertained on the subject of Light, before the time of Sir Isaac Newton, were vague and unsatisfactory. It was shown by that eminent philosopher, that a ray of sunlight was not *homogeneous*, as had been supposed, but consisted of several rays of vivid colours, united and intermingled.

This fact may be demonstrated by throwing a pencil of Sunlight upon one angle of a *prism*, and receiving the oblong image, so formed, upon a white screen.

The space illuminated and coloured by a pencil of rays

analyzed in this way is called " the Solar Spectrum." The action of a prism in decomposing white light will be more fully explained in the next Section. At present we notice only that seven principal colours may be distinguished in

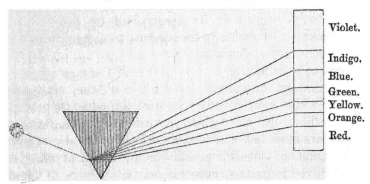

Violet.

Indigo.
Blue.
Green.
Yellow.
Orange.
Red.

the Solar Spectrum, viz. red, orange, yellow, green, blue, indigo, and violet. Sir David Brewster has made observations which lead him to suppose that the *primary* colours are in reality but three in number, viz. red, yellow, and blue, and that the others are *compound*, being produced by two or more of these overlapping each other; thus the red and yellow spaces intermingled constitute *orange;* the yellow and blue spaces, *green.*

The composition of white light from the seven prismatic colours may be roughly proved by painting them on the face of a wheel, and causing it to rotate rapidly; this blends them together, and a sort of greyish-white is the result. The white is imperfect, because the colours employed cannot possibly be obtained of the proper tints or laid on in the exact proportions.

The decomposition of light is effected in other ways besides that already given :—

First, by *reflection* from the surfaces of coloured bodies. All substances throw off rays of light, which impinge upon the retina of the eye and produce the phenomena of vision. *Colour* is caused by a *portion only,* and not the whole, of the elementary rays, being projected in this way. Surfaces termed *white* reflect all the rays; coloured surfaces absorb some and reflect others: thus *red* substances reflect only red rays, *yellow* substances, yellow rays, etc., the ray which is reflected in all cases deciding the colour of the substance.

Secondly, light may be decomposed by *transmission* through media which are transparent to certain rays, but opaque to others.

Ordinary transparent glass allows all the rays constituting white light to pass; but by the addition of certain metallic oxides to it whilst in a state of fusion, its properties are modified, and it becomes *coloured*. Glass stained by Oxide of Cobalt is permeable only to blue rays. Oxide of Silver imparts a pure yellow tint; Oxide of Gold or Suboxide of Copper a ruby red, etc.

DIVISION OF THE ELEMENTARY RAYS OF WHITE LIGHT INTO LUMINOUS, HEAT-PRODUCING, AND CHEMICAL RAYS.

The agency of Light produces a variety of distinct effects upon the bodies which surround us. These may be classed together as the properties of light. They are of three kinds—the phenomena of colour and vision, of heat, and of chemical action.

By resolving white light into its constituent rays, we find that these properties are associated each one with certain of the elementary colours.

The *yellow* is decidedly the most luminous ray. On examining the Solar Spectrum, it is seen that the brightest part is that occupied by the yellow, and that the light diminishes rapidly on either side. So again, rooms glazed with yellow glass always appear abundantly illuminated, whilst the effect of red or blue glass is dark and sombre. The yellow colour therefore constitutes that portion of white light by which surrounding objects are rendered visible; it is essentially the *visual* ray.

The *heating properties* of the sunlight reside principally in the red ray, as is shown by the expansion of a mercurial thermometer placed in that part of the spectrum.

The chemical action of light corresponds more to the indigo and violet rays, and is wanting, as regards its influence upon Iodide of Silver, both in the red and yellow. Strictly speaking however it cannot be localized in either of the coloured spaces, as will be more fully shown in the Fourth Section of this Chapter, to which the reader is referred.

SECTION II.

The Refraction of Light.

A ray of light, in its passage through any transparent medium, travels in a straight line as long as the density of the medium continues unchanged. But if the density varies, becoming either greater or less, then the ray is *refracted*, or bent out of the course which it originally pursued. The degree to which the refraction or bending takes place depends upon the nature of the new medium, and in particular upon its *density* as compared with that of the medium which the ray had previously traversed. Hence Water refracts light more powerfully than Air, and Glass more so than Water.

The following diagram illustrates the refraction of a ray of light.

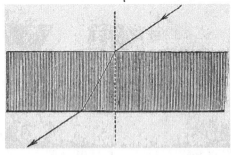

The dotted line is drawn perpendicularly to the surface,

E

and it is seen that the ray of light on entering is bent *towards* this line. On emerging, on the other hand, it is bent to an equal extent *away from the perpendicular,* so that it proceeds in a course parallel to, but not coincident with, its original direction. If we suppose the new medium, in place of being more dense than the old, to be *less dense,* then the conditions are exactly reversed,—the ray is bent away from the perpendicular on entering, and towards it on leaving.

It must be observed that the laws of refraction apply only to rays of light which fall upon the medium *at an angle:* if they enter perpendicularly—in the direction of the dotted lines in the last figure—they pass straight through without suffering refraction.

Notice also, that it is *at the surfaces of bodies* that the deflecting power acts. The ray is bent on entering, and bent again on leaving; but whilst within the medium it continues in a straight line. Hence it is evident that by variously modifying the surfaces of refractive media the rays of light may be diverted almost at pleasure. This will be rendered clear by a few simple diagrams.

In the figures given below, and in the following page, the dotted lines represent perpendiculars to the surface at the point where the ray falls, and it is seen that the usual

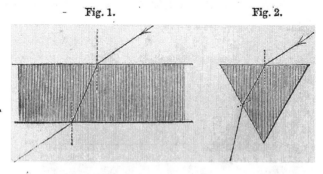

Fig. 1.   Fig. 2.

law of bending *towards* the perpendicular on entering, and *away* from it on leaving the dense medium, is in each case correctly observed.

Fig. 2, termed a prism, bends the ray permanently to one side; fig. 3, consisting of two prisms placed base to base, causes rays before parallel to meet in a point; and conversely, fig. 4, having prisms placed edge to edge, diverts them further asunder.

Fig. 3.   Fig. 4.

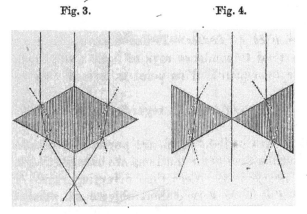

The various forms of Lenses.—The phenomena of the refraction of light are seen in the case of *curved* surfaces in the same manner as with those which are plane.

Glasses ground of a curvilinear form are termed *Lenses.* The following are examples.

Fig. 1.   Fig. 2.   Fig. 3.

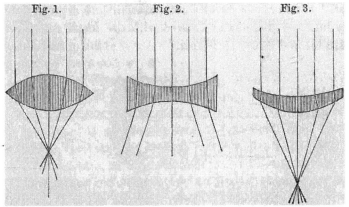

Fig. 1 is a biconvex lens; fig. 2, a biconcave lens; and fig. 3, a *meniscus* lens.

As far as regards their refractive powers, such figures may be represented, nearly, by others bound by straight lines, and thus it becomes evident that a biconvex lens tends to condense rays of light to a point, and a biconcave to scatter them. A meniscus combines both actions, but the rays are eventually bent together, the convex curve of a meniscus lens being always greater than the concave.

The Foci of Lenses.—It has been shown that convex lenses tend to condense rays of light and bring them together to a point. This point is termed "the focus" of the Lens.

The following laws as regards the focus may be laid down:—

That rays of light which are pursuing a parallel course at the time they enter the Lens are brought to a focus at a point nearer to the Lens than diverging rays. The rays proceeding from very distant objects are parallel; those from objects near at hand diverge. The sun's rays are always parallel, and the divergence of the others becomes greater as the distance from the Lens is less.

The focus of a Lens for parallel rays is termed the "principal focus," and is not subject to variation; this is the point referred to when the *focal length* of a Lens is spoken of. When the rays are not parallel, but diverge from a point, that point is associated with the focus, and the two are termed "conjugate foci."

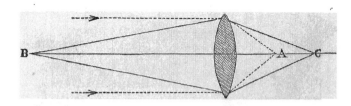

In the above diagram A is the principal focus, and B and C are conjugate foci. Any object placed at B has its focus at C, and conversely when placed at C it is in focus at B.

Therefore, although the principal focus of a Lens (as determined by the degree of its convexity) is always the same,
yet the focus for objects near at hand varies, being longer
as they are brought closer to the Lens.

Formation of a Luminous Image by a Lens.—As the
rays of light proceeding from a *point* are brought to a
focus by means of a Lens, so are they when they proceed
from an object, and in that case *an image of the object* is
the result.

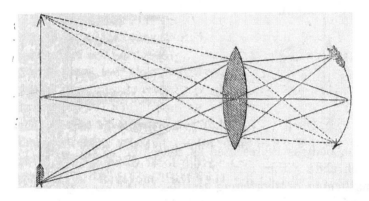

The above figure illustrates this. The *size* of the image
varies with the distance of the arrow from the glass—
being larger and formed at a point further from the Lens
as the object is brought nearer. The refracting power of
the Lens also influences the result—lenses of short focal
length, *i. e.* more convex, giving a smaller image.

In order that the course pursued by pencils of rays proceeding from an object may be easily traced, the lines from
the barb of the arrow in the last figure are *dotted*. Observe
that the object is necessarily *inverted*, and also that those
rays which traverse the central point of the Lens, or the
centre of the *axis*, as it is termed, are not bent away, but
pursue a course either coincident with, or parallel to, the
original, as in the case of refracting media with parallel
surfaces.

SECTION III.

The Photographic Camera.

The Photographic Camera is in its essential nature an extremely simple instrument. It consists merely of a *dark chamber*, having an aperture in front in which a Lens is inserted. The accompanying figure shows the simplest form of Camera.

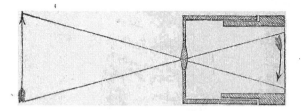

The body is represented as consisting of two portions which slide within each other; but the same object of lengthening or shortening the focal distance may be attained by making the Lens itself movable. A luminous image of any object placed in front of the Camera is formed by means of the Lens, and received upon a surface of ground glass at the back part of the instrument. When the Camera is required for use, the object is *focussed* upon the ground glass, which is then removed, and a slide containing the sensitive layer inserted in its place.

The luminous image, as formed upon the ground glass, is termed the "Field" of the Camera; it is spoken of as being flat or curved, sharp or indistinct, etc. These and other peculiarities which depend upon the construction of the Lens will now be explained.

Chromatic Aberration of Lenses.—The outside of a biconvex lens is strictly comparable with the sharp edge of a *prism*, and therefore necessarily produces decomposition in the white light which passes through it.

The action of a prism in separating white light into its constituent rays may be simply explained;—all the co-

loured rays are refrangible, but not to the same extent.
The indigo and violet are more so than the yellow and red,
and consequently they are separated from them, and occupy a higher position in the Spectrum. (See the diagram
at p. 47.)

A little reflection will show that in consequence of this
unequal refrangibility of the coloured rays, white light
must invariably be decomposed on entering any dense
medium.   This is indeed the case; but if the surfaces of
the medium *are parallel to each other* the effect is not
seen, because the rays recombine on their emergence,
being bent to the same extent in the opposite direction.
Hence light is transmitted colourless through an ordinary
pane of glass, but yields the tints of the Spectrum in its
passage through a prism or a lens, where the two surfaces
are inclined to each other at an acute angle.

Chromatic aberration is corrected by combining two
lenses cut from varieties of glass which differ in their
power of separating the coloured rays.   These are the
dense flint-glass containing Oxide of Lead, and the light
crown-glass.   Of the two lenses, the one is *biconvex*, and
the other *biconcave;* so that when fitted together they
produce a compound Achromatic lens of a meniscus form,
thus :—

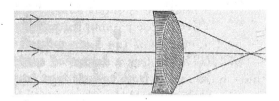

The first Lens in this figure is the flint- and the second
the crown-glass.   Of the two the biconvex is the most
powerful, so as to overcome the other, and produce a total
of refraction to the required extent.   Each of the Lenses
produces a spectrum of a different length; and the effect

of passing the rays through both, is, by overlapping the coloured spaces, to unite the complementary tints, and to form again white light.

Spherical Aberration of Lenses.—The field of a Camera is not often equally sharp and distinct at every part. If the centre be rendered clear and well defined, the outside is misty; whilst, by slightly altering the position of the ground glass, so as to define the outside portion sharply, the centre is thrown out of focus. Opticians express this by saying that there is a want of proper *flatness* of field; two causes may be mentioned as concurring to produce it.

The first is " spherical aberration," by which is meant the property possessed by Lenses which are segments of *spheres*, of refracting rays of light *unequally* at different parts of their surfaces. The following diagram shows this :—

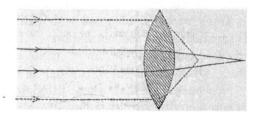

Observe that the dotted lines which fall upon the circumference of the Lens are brought to a focus at a point nearer to the Lens than those passing through the centre; in other words, the outside of the Lens refracts light the most powerfully. This causes a degree of confusion and indistinctness in the image, from various rays crossing, and interfering with, each other.

Spherical aberration may be avoided by increasing the *convexity* of the centre part of the Lens, so as to add to its refracting power at that particular point. The surface is then no longer a segment of a sphere, but of an *ellipse*, and refracts light more equally. The difficulty of grinding Lenses to an elliptical form however is so great, that the spherical Lens is still used, the aberration being corrected in other ways.

A second cause interfering with the distinctness of the outer portions of the image in the Camera is the *obliquity* of some rays proceeding from the object; in consequence of which the image has a curved form, with the concavity inwards, as may be seen by referring to the figure given at page 53. The following diagram is meant to explain curvature of the image.

The centre line running at right angles to the general direction of the Lens is the *axis;* an imaginary line, on which the Lens may be said to rotate as a wheel turns on its axle. The lines A A represent rays of light falling parallel to the axis; and the dotted lines, others which have an *oblique* direction; B and C show the points at

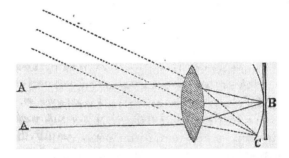

which the two foci are formed. Observe that these points, although equidistant from the centre of the Lens, do not fall in the same vertical plane, and therefore they cannot both be received distinct upon the ground glass of the Camera, which would occupy the position of the perpendicular double line in the diagram. Hence it is that with most lenses, when the centre of the field has been focussed, the glass must be shifted forwards a little to define the outside sharply.

The Use of Stops in Lenses.—Curvature of the image and indistinctness of outline from spherical aberration are both remedied to a great extent by fixing in front of the

Lens a diaphragm having a small central aperture. The diagram gives a sectional view of a Lens with a "stop"

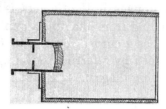

attached; the exact position it should occupy with reference to the Lens is a point of importance, and influences the flatness of the field.

By using a diaphragm the quantity of light admitted into the Camera is diminished in proportion to the size of the aperture. The image is therefore less brilliant, and a longer exposure of the sensitive plate is required. In other respects however the result is improved; the spherical aberration is lessened by cutting off the outside of the Lens, and a portion of the oblique rays being intercepted, the focus of the remainder is lengthened out, and the image is rendered flatter, and improved in distinctness. Hence also, when a small stop is affixed to a Lens, a variety of objects, situated at different distances, are all in focus at once; whereas, with the full aperture of the Lens, objects near at hand cannot be rendered distinct upon the ground glass at the same time with distant objects, or *vice versâ.*

The Double or Portrait Combination of Achromatic Lenses.—The brightness of illumination of an image formed by a Lens is in proportion to the *diameter* of the Lens, that is, to the size of the aperture by which the Light is admitted. The *clearness or distinctness of outline* however is independent of this, being improved by using a stop, which lessens the diameter.

The Portrait combination of Lenses is constructed to ensure rapidity of action by admitting a large volume of light. The following diagram gives a sectional view.

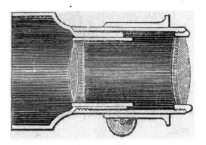

In this combination the front Lens is an Achromatic plano-convex, with the convex side turned toward the object; and the second, which takes up the rays and refracts them further, is a compound Biconvex Lens; there are therefore in all four distinct glasses concerned in forming the image, which may appear at first to be an unnecessarily complex arrangement. It is found however that a good result cannot be secured by using a single Lens, when a "stop" is inadmissible. By combining two glasses of different curves, the aberrations of one correct those of the other to a great extent, and the field is both flatter and more distinct than in the case of an Achromatic Meniscus employed without a diaphragm.

The manufacture of Portrait Lenses is a point of great difficulty, the glasses requiring to be ground with extreme care, in order to avoid *distortion* of the image : hence the most rapid Portrait Lenses, having large aperture and short focus, are often useless unless purchased of a good maker.

The Variation between the Visual and Actinic Foci in Lenses.—The same causes which produce chromatic aberration in a Lens, tend also to separate the chemical from the visual focus.

The violet and indigo rays are more strongly bent in than the yellow, and still more than the red; consequently the focus for each of those colours is at a different point. The following diagram shows this.

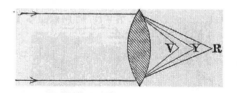

V represents the focus of the violet ray, Y of the yellow, and R of the red.

Hence, as the chemical action corresponds more to the violet, the most marked actinic effect would be produced at V. The luminous portion of the spectrum however is *the yellow*, consequently the visual focus is at Y.

Photographers have long recognized this point; and therefore, with ordinary Lenses, not corrected for colour, rules are laid down as to the exact distance which the sensitive plate should be shifted away from the visual focus in order to obtain the greatest amount of distinctness of outline in the image impressed by chemical action.

These rules do not apply to the Achromatic Lenses recently described. The coloured rays being in that case bent together again and reunited, the two foci also nearly correspond. By a little further correction to a point higher in the Spectrum, they are made to do so perfectly.

SECTION IV.

On the Photographic Action of Coloured Light.

It has already been mentioned in the First Section of this Chapter that certain of the elementary colours of white light, viz. the violet and indigo, are peculiarly active in decomposing the Photographic Salts of Silver; but there are some points of importance relating to the same subject which require a further notice.

The term "actinism" (Gr. ἀκτίς, a ray or flash) has been proposed as convenient to designate the property possessed by light of producing chemical change; the rays to which the effect is especially due being known as *actinic* rays.

If the pure Solar Spectrum formed by prismatic analysis in the manner represented at page 47 be allowed to impinge upon a prepared sensitive surface of Iodide of Silver, the latent image being subsequently developed by a reducing agent, the effect produced will be something similar to that represented in the following diagram :—

Fig. 1.   Fig. 2.

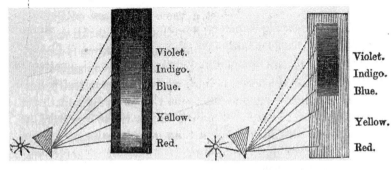

Fig. 1 shows the visible spectrum as it appears to the eye; the brightest part being in the yellow space, and the light gradually shading off until it ceases to be seen. Fig. 2 represents the chemical effect produced by throwing the Spectrum upon Iodide of Silver. Observe that the darkening characteristic of chemical action is most evident in the upper spaces, where the *light* is feeble, and is altogether absent at the point corresponding to the bright yellow spot of the visible spectrum. The actinic and luminous spectra are therefore totally distinct from each other, and the word "Photography," which signifies the process of taking pictures by *light*, is in reality inaccurate.

To those who have not the opportunity of working with the Solar Spectrum, the following experiments will be useful in illustrating the Photographic value of coloured light.

Experiment I.—Take a sheet of sensitive paper prepared with Chloride of Silver, and lay upon it strips of blue, yellow, and red glass. On exposure to the sun's rays for a few minutes, the part beneath the blue glass darkens rapidly, whilst that covered by the red and yellow glass is perfectly protected. This result is the more striking from the extreme *transparency* of the yellow glass, giving the idea that the Chloride would certainly be blackened first at that point. On the other hand, the blue glass appears very dark, and effectually conceals the tissue of the paper from view.

Experiment II.—Select a vase of flowers of different shades of scarlet, blue, and yellow, and make a Photographic copy of them, by development, upon Iodide of Silver. The blue tints will be found to act most violently upon the sensitive compound, whilst the reds and yellows are scarcely visible; were it not that it is difficult to procure in nature pure and homogeneous tints, free from admixture with other colours, they would make no impression whatever upon the plate.

In exemplifying further the importance of distinguishing between visual and actinic rays of light, we may observe that if the two were in all respects the same, Photography must cease to exist as an Art. It would be impossible to make use of the more sensitive chemical preparations from the difficulties which would attend the previous preparation and subsequent development of the plates. These operations are now conducted in what is termed a *dark* room; but it is dark only in a *Photographic* sense, being illuminated by means of yellow light, which, whilst it enables the operator easily to watch the progress of the work, produces no injurious effect upon the sensitive surfaces. If the windows of the room were glazed with *blue* in place of yellow glass, then it would be strictly a "dark room," but one altogether unfitted for the purpose intended.

Another point connected with the same subject and

worthy of note is—the extent to which the sensibility of the Photographic compounds is influenced by atmospheric conditions not visibly interfering with the *brightness* of the light. It is natural to suppose that those days on which the sun's rays are the most powerful would be the best for rapid impression, but such is not by any means the case. If the light is at all of a yellow cast, however bright it may be, its actinic power will be small.

It will also be often observed in working towards the evening, that a sudden diminution of sensibility in the plates begins to be perceptible at a time when but little difference can be detected in the brilliancy of the light; the setting sun has sunk behind a golden cloud, and all chemical action is soon at an end.

In the same manner is explained the difficulty of obtaining Photographs in the glowing light of tropical climates; the superiority of the early months of spring over those of the midsummer; of the morning sun to that of the afternoon, etc. April and May are usually considered the best months for rapid impression in this country; but the light continues good until the end of July. In August and September a longer exposure of the plates will be required.

THE SUPERIOR SENSIBILITY OF BROMIDE OF SILVER TO COLOURED LIGHT.

In copying the Solar Spectrum alternately upon a surface of Iodide and Bromide of Silver, we notice a difference in the Photographic properties of these two salts. The latter is affected more extensively, to a point lower in the spectrum, than the former. In the case of the Iodide of Silver, the action ceases in the Blue space; but with the Bromide it reaches to the Green. This is shown in the following diagrams, which are drawn from the observations of Mr. Crookes ('Photographic Journal,' vol. i. p. 100):—

Fig. 1. Fig. 2. Fig. 3.

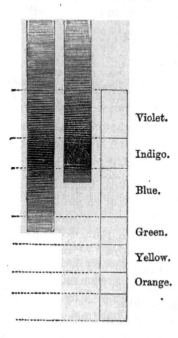

Violet.

Indigo.

Blue.

Green.

Yellow.

Orange.

Fig. 1 represents the chemical spectrum on Bromide of Silver; fig. 2, the same upon Iodide of Silver; and fig. 3, the visible spectrum.

It might perhaps be supposed that the superior sensibility of the Bromide of Silver to green rays of light would render that salt useful to the Photographer in copying landscape scenery; and indeed it is the opinion of many that, in the *Calotype* paper process, the dark colour of foliage is better rendered by a mixture of Bromide and Iodide of Silver than by the latter salt alone. This however cannot depend upon the greater sensibility of the Bromide to coloured light, as may easily be proved.—

The diagrams given above are shaded to represent nearly, the relative intensity of the chemical action exerted by the rays at different points of the spectrum; and on referring to them it will be seen that the maximum point of blackness is in the indigo and violet space, the action being more feeble in the blue space lower down; there are also highly refrangible rays extending upwards far beyond the visible colours, and these invisible rays are actively concerned in the formation of the image.

It is evident therefore that the amount of effect produced by a pure green, or even a light blue tint, upon a surface of Bromide of Silver is very small as compared with that of an indigo or violet; and hence, as in copying natural objects radiations of all kinds are present at the same time, the green tints have not time to act before the image is impressed by the more refrangible rays.

Sir John Herschel proposed to render coloured light more available in Photography by separating the actinic rays of high refrangibility, and working only with those which correspond to the blue and green spaces in the spectrum. This may be done by placing in front of the Camera a vertical glass trough containing a solution of Sulphate of Quinine. Professor Stokes has shown that this liquid possesses curious properties. In transmitting rays of light it *modifies* them so that they emerge *of lower refrangibility*, and incapable of producing the same actinic effect. Sulphate of Quinine is, if we may use the term, *opaque* to all actinic rays higher than the blue-coloured space. The proposition of Sir John Herschel above referred to was therefore to employ a bath of Sulphate of Quinine, and having eliminated the actinic rays of high refrangibility, to work upon Bromide of Silver with those corresponding to the lower-coloured spaces. In this way he conceived that a more natural effect might be obtained.

If Photographic compounds should be discovered of greater sensibility than any we at present possess, the use of the Quinine bath will perhaps be adopted; but at present we trust to the superior intensity of the invisible rays for the formation of the image, and hence the employment of Bromide of Silver is less strongly indicated.

These remarks apply to Photographs taken by sunlight.

F

Mr. Crookes states that in working with artificial light, such as gas or camphine, the case is different. Actinic rays of high refrangibility are comparatively wanting in gas-light, the great bulk of the Photographic rays being found to lie within the limits of the visible spectrum, and consequently acting more energetically upon Bromide than on Iodide of Silver.

Explanation of the mode in which Coloured Objects impress the Sensitive Film.—The fact of which we have been speaking, viz. that the natural colours are not always correctly represented in photography, is often urged in depreciation of the art,—"when lights are represented by shadows," it is said, "how can a truthful picture be expected?" The insensitiveness of Iodide of Silver to the colours occupying the lower portion of the spectrum would indeed present an insuperable difficulty *if the tints of Nature were pure and homogeneous:* such however is not the case. Even the most sombre colours are accompanied by scattered rays of white light in quantity amply sufficient to affect the sensitive film.

This is especially seen when the coloured body *possesses a good reflecting surface;* and hence some varieties of foliage, as for instance the Ivy, with its smooth and polished leaf, are more easily photographed than others. So again with regard to drapery in the department of portraiture— it is necessary to attend not only to the colour, but also *to the material of which it is composed.* Silks and satins are favourable, as reflecting much light, whilst velvets and coarse stuffs of all kinds, if at all dark, produce very little effect upon the sensitive film.

SECTION V.

On Binocular Vision and the Stereoscope.

An object is said to be "stereoscopic" (στερεος, solid, and σκοπεω, I see) when it stands out in relief, and gives to the eye the impression of solidity.

This subject was first explained by Professor Wheatstone in a memoir on binocular vision, published in the 'Philosophical Transactions' for 1838; in which he shows that solid bodies project different perspective figures upon each retina, and that the illusion of solidity may be artificially produced by means of the " Stereoscope."

The phenomena of binocular vision may be simply sketched as follows:—If a cube, or a small box of an oblong form, be placed at a short distance in front of the observer, and viewed attentively with the right and left eye separately and in succession, it will be found that the figure perceived in the two cases is different; that each eye sees more of one side of the box, and less of the other; and that in neither instance is the effect exactly the same as that given by the two eyes employed conjointly.

A silver pencil-case, or a pen-holder, may be used to illustrate the same fact. It should be held at about six or eight inches distant from the root of the nose, and quite at right angles to the face, so that the length of the pencil is concealed by the point. Then, whilst it remains fixed in this position, the left and right eye are to be alternately closed: in each case a portion of the opposite side of the pencil will be rendered visible.

Fig. 1.   Fig. 2.

The preceding diagrams exhibit the appearance of a bust as seen by each eye successively.

Observe that the second figure, which represents the impression received by the right eye, is more of a full face than fig. 1, which, being viewed from a point removed a little to the left, partakes of the character of a profile.

The human eyes are placed about $2\frac{1}{2}$ inches, or from that to $2\frac{5}{8}$ inches, asunder; hence it follows that, the points of sight being separated, a *dissimilar* image of a solid object is formed by each eye. We do not however see *two* images, but a single one, which is stereoscopic.

In looking at a picture painted on a flat surface the case is different: the eyes, as before, form two images, but these images are in every respect similar; consequently the impression of solidity is wanting. A single picture, therefore, cannot be made to appear stereoscopic. To convey the illusion *two* pictures must be employed, the one being a right and the other a left perspective projection of the object. The pictures must also be so arranged, that each is presented to its own eye, and that the two appear to proceed from the same spot.

The reflecting stereoscope, employed to effect this, forms *luminous images* of the binocular pictures, and throws these images together, so that, on looking into the instrument, only a single image is seen, in a central position. It should, however, be understood, that no optical arrangement of any kind is indispensably required, since it is quite possible, with a little effort, to combine the two images by the unaided organs of vision. The following diagram will make this obvious :—

The circles A and B represent two wafers, which are stuck on paper at a distance of about three inches from each other. They are then viewed by *squinting* strongly, or turning the eyes inwards towards the nose, until the right eye looks at the left wafer, and the left eye at the right wafer. Each wafer will then appear to become double, four images being seen, the two central of which will gradually approach each other until they coalesce. Stereoscopic pictures, properly arranged, may be examined in the same manner; and it will be found that the resultant solid image is formed midway, at a point where two lines, drawn across from the eyes to the pictures, cut one another. The experiment here mentioned is sometimes a painful one, and cannot easily be made if the eyes are not of equal strength; but it will serve to show that the essential principle resides in the binocular representation of the object, and not in the instrument employed to view it.

In Mr. Wheatstone's reflecting Stereoscope *mirrors* are used. The principle of the instrument is as follows:— objects placed in front of a mirror have their reflected images apparently *behind* the mirror. By arranging two mirrors at a certain inclination to each other, the images of the double picture may be made to approach until they coalesce, and the eye perceives a single one only. The following diagram will explain this.

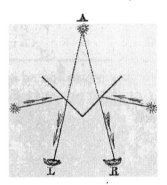

The rays proceeding from the star on either side pass in the direction of the arrows, being thrown off from the mirror (represented by the thick black line) and entering the eyes at R and L. The reflected images appear behind the mirror, uniting at the point A.

The reflecting Stereoscope is adapted principally for viewing large pictures. It is a very perfect instrument, and admits of a variety of adjustments, by which the apparent size and distance of the Stereoscopic image may be varied almost at pleasure.

The "lenticular" Stereoscope of Sir David Brewster is a more portable form of apparatus. A sectional view is given in the diagram.

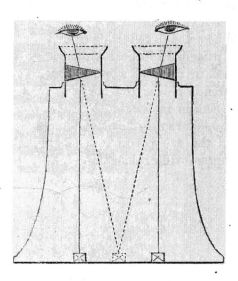

The brass tubes to which the eyes of the observer are applied contain each a semi-lens, formed by dividing a common lens through the centre and cutting each half into a circular form (fig. 1 in the following page). The half-lens viewed in section (fig. 2) is therefore of a prismatic shape, and when placed with its sharp edge as in the diagram above, alters the direction of the rays of light proceeding from the picture, bending them outwards or away from the centre, so that in accordance with well-known

Fig. 1.

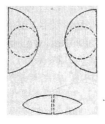

Fig. 2.

optical laws they *appear* to come in the direction of the
dotted lines in the diagram (in the last page), and the two
images coalesce at their point of junction. In the instrument as it is often sold, one of the lenses is made movable, and by turning it round with the finger and thumb it
will be seen that the positions of the images may be shifted
at pleasure.

Rules for taking Binocular Photographs.—In viewing
very distant objects with the eyes, the images formed on
the retinæ are not sufficiently dissimilar to produce a very
Stereoscopic effect; hence it is often required, in taking binocular pictures, to separate the Cameras more widely than
the two eyes are separated, in order to give a sufficient appearance of relief. Mr. Wheatstone's original directions
were, to allow about one foot of separation for each twenty-
five feet of distance, but considerable latitude may be
permitted.

If the Cameras be not placed far enough apart, the dimensions of the stereoscopic image from before backwards
will be too small,—statues looking like bas-reliefs, and the
circular trunks of trees appearing oval, with the long diameter transverse. On the other hand, when the separation
is too wide, the reverse obtains,—objects for instance
which are square, assuming an oblong shape pointing towards the observer.

To understand the cause of this, the following law in
optics should be studied:—"The *distance* of objects is

estimated by the extent to which the axes of the eyes must be converged to view them." If we have to turn our eyes strongly inwards, we judge the object to be near; but if the eyes remain nearly parallel, we suppose it to be distant.

The above figures represent six-sided truncated pyramids, each with its apex towards the observer, the centres of the two smaller interior hexagons being more widely separated than those of the larger exterior ones. By converging the eyes upon them so as to unite the central images in the manner represented in page 68 a greater amount of convergence will be required to bring together the two summits than the bases, and hence the summits will appear the nearest to the eye; that is to say, the resultant central figure will acquire the additional dimension of *height*, and appear as a solid cone, standing perpendicularly upon its base: further, the more widely the summits are separated in relation to the bases, the taller will the cone be, although a greater effort will be required to coalesce the figures.

Binocular Photographs taken with too much separation of the Cameras, are distorted from a similar cause,—so strong a convergence being required to unite them that certain parts of the picture appear to approach near to the eye; and the depth of the solid image is increased.

This effect is most observable when the picture embraces a variety of objects, situated in different planes. In the case of views which are quite distant, no near objects being admitted, the Cameras may be placed with especial reference to them, even as far as twelve feet apart, without producing distortion.

It is sometimes observable, in looking at Stereoscopic

pictures, that they convey an erroneous impression of the real size and distance of the object. For instance, in using the large reflecting Stereoscope, if, when the adjustments have been made and the images properly united, the two pictures be moved slowly forward, the eyes remaining fixed upon the mirrors, the Stereoscopic image will gradually change its character, the various objects it embraces appearing to become diminished in size, and approaching near to the observer: whilst if the pictures be pushed *backwards*, the image will enlarge and recede to a distance. So, again, if an ordinary slide for the lenticular Stereoscope be divided in the centre, and, looking into the instrument until the images coalesce, the two halves be slowly separated from each other, the solid picture will seem to become larger and to recede from the eye.

It is easy to understand the cause of this. When the pictures in the reflecting Stereoscope are moved *forwards*, the convergence of the optic axes is increased: the image therefore appears *nearer*, in accordance with the last-mentioned law. But to convey the impression of nearness is equivalent to an apparent diminution in size, for we judge of the dimensions of a body very much in relation to its supposed distance. Of two figures, for instance, appearing of the same height, one known to be a hundred yards off might be considered colossal, whilst the other, obviously near at hand, would be viewed as a statuette.

These facts, with others not mentioned, are of great interest and importance, but their further consideration does not fall within the bounds originally prescribed to us. The practical details of Stereoscopic Photography have been arranged in a distinct Section, and will be found included in the Second Part of the Work.*

* For a more full and detailed explanation of the Stereoscopic phenomena, see an abstract of Professor Tyndall's lectures in the third volume of the 'Photographic Journal.'

CHAPTER VI.

THE PHOTOGRAPHIC PROPERTIES OF IODIDE OF SILVER UPON COLLODION.

In the preceding part of this Work the physical and chemical properties of Chloride and Iodide of Silver have been described, with the changes which they experience by the action of Light. Nothing however has been said of the surface used to support the Iodide of Silver, and to expose it in a finely divided state to the influence of the actinic radiations. This omission will now be supplied, and the use of Collodion will engage our attention.

The sensibility of Iodide of Silver upon Collodion is greatly superior to that of the same salt employed in conjunction with any other vehicle at present known. Hence the Collodio-Iodide film will supersede the paper and Albumen processes in all cases where objects liable to move are to be copied. The causes of this superior sensitiveness, as far as ascertained, may be referred to the state of *loose coagulation* of a Collodion film and other particulars presently to be noticed. It must however be allowed that there are yet some points affecting the sensitiveness of Iodide of Silver, both mechanical and chemical, of the exact nature of which we are ignorant.

The present Chapter may be divided into four Sections: —the nature of Collodion; the chemistry of the Nitrate Bath; the causes affecting the formation and development of the Image upon Collodion; the various irregularities in the development of the Image.

SECTION I.

Collodion.

Collodion (so named from the Greek word κολλάω, *to stick*) is a glutinous, transparent fluid, procured, as generally said, by dissolving Gun Cotton in Ether. It was originally used for surgical purposes only, being smeared over wounds and raw surfaces, to preserve them from contact with the air by the tough film which it leaves on evaporation. Photographers employ it to support a delicate film of Iodide of Silver upon the surface of a smooth glass plate.

Two elements enter into the composition of Collodion: first, the Gun Cotton; second, the fluids used to dissolve it. Each of these will be treated in succession.

CHEMISTRY OF PYROXYLINE.

Gun Cotton or *Pyroxyline* is Cotton or Paper which has been altered in composition and properties by treatment with strong acids.

Both Cotton and Paper are, chemically, the same. They consist of fibres which are found on analysis to have a constant composition, containing three elementary bodies, Carbon, Hydrogen, and Oxygen, united together in fixed proportions. To this combination the term *Lignine* or *Cellulose** has been applied.

Cellulose is a definite chemical compound, in the same sense as Starch or Sugar, and consequently, when treated with various reagents, it exhibits properties peculiar to itself. It is insoluble in most liquids, such as Water, Alcohol, Ether, etc., and also in dilute acids; but when

* Lignine and Cellulose are not precisely identical substances. The latter is the material composing the cell-wall; the former, the contained matter in the cell.

acted upon by *Nitric Acid* of a certain strength it liquefies and dissolves.

It has been already shown (p. 12) that when a body dissolves in Nitric Acid the solution is not usually of the same nature as an aqueous solution; and so in this case—the Nitric Acid imparts Oxygen first to the Cotton, and afterwards dissolves it.

Preparation of Pyroxyline.—If, instead of treating Cotton with Nitric Acid, a mixture of Nitric and Sulphuric Acids in certain proportions be used, the effect is peculiar. The fibres contract slightly, but undergo no other visible alteration. Hence we are at first disposed to think the mixed Acids ineffectual. This idea however is not correct, since on making the experiment the *properties* of the cotton are found to be changed. Its weight has increased by more than one-half; it has become soluble in various liquids, such as Acetic Ether, Ether and Alcohol, etc., and, what is more remarkable, it no longer burns in the air quietly, but *explodes* on the application of flame with greater or less violence.

This change of properties clearly shows, that although the fibrous structure of the material is unaffected, it is no longer the same substance, and consequently chemists have assigned it a different name, viz. *Pyroxyline*.

To produce the peculiar change by which Cotton is converted into Pyroxyline, both Nitric and Sulphuric Acids are, as a rule, required; but of the two the former is the most important. On analyzing Pyroxyline, Nitric Acid, or a body analogous to it, is detected in considerable quantity, but not Sulphuric Acid. The latter Acid, in fact, serves but a temporary purpose, viz. *to prevent the Nitric Acid from dissolving the Pyroxyline*, which it would be liable to do if employed alone. The Sulphuric Acid prevents the solution by removing water from the Nitric Acid, and so producing a higher degree of concentration; Pyroxyline, although soluble in a dilute, is not so in the strong Acid, and hence it is preserved.

The property possessed by Oil of Vitriol of removing water from other bodies, is one with which it is well to be acquainted. A simple experiment will serve to illustrate it. Let a small vessel of any kind be filled to about two-thirds with Oil of Vitriol, and set aside for a few days; at the end of that time, and especially if the atmosphere be damp, it will have absorbed sufficient moisture to cause it to flow over the edge.

Now even the strongest reagents employed in chemistry contain, almost invariably, water in greater or lesser quantity. Pure Anhydrous Nitric Acid is a white, solid substance; Hydrochloric Acid is a gas: and the liquids sold under those names are merely *solutions*. The effect then of mixing strong Oil of Vitriol with aqueous Nitric Acid is to remove water in proportion to the amount used, and to produce a liquid containing Nitric Acid in a high state of concentration, and Sulphuric Acid more or less diluted. This liquid is the *Nitro-Sulphuric Acid* employed in the preparation of Pyroxyline.

Various forms of Pyroxyline.—Very soon after the first announcement of the discovery of Pyroxyline, most animated discussions arose amongst chemists with regard to its solubility and general properties. Some spoke of a "solution of Gun Cotton in Ether;" whilst others denied its solubility in that menstruum; a third class, by following the process described, obtained a substance which was not explosive, and therefore could scarcely be termed Gun Cotton.

On further investigations some of these anomalies were cleared up, and it was found that there were *varieties* of Pyroxyline, depending mainly upon the degree of strength of the Nitro-Sulphuric Acid employed in the preparation. Still the subject was obscure until the publication of researches by Mr. E. A. Hadow. These investigations, conducted in the Laboratory of King's College, London, were published in the Journal of the Chemical Society. Constant reference will be made to them in the following remarks.

We notice—*first*, the chemical constitution of Pyroxyline; *secondly*, its varieties; and *thirdly*, the means adopted to procure a Nitro-Sulphuric Acid of the proper strength.

a. *Constitution of Pyroxline.*—Pyroxyline has been sometimes spoken of as a Salt of Nitric Acid, a *Nitrate of Lignine.* This view however is erroneous, since it can be shown that the substance present is not Nitric Acid, although analogous to it. It is the Peroxide of Nitrogen, which is intermediate in composition between Nitrous Acid (NO_3) and Nitric Acid (NO_5). Peroxide of Nitrogen (NO_4) is a gaseous body of a dark red colour; it possesses no acid properties, and is incapable of forming a class of salts. In order to understand in what state this body is combined with cotton fibre to form Pyroxyline, it will be necessary to digress for a short time.

Law of Substitution.—By the careful study of the action of Chlorine, and of Nitric Acid, upon various organic substances, a remarkable series of compounds has been discovered, containing a portion of Chlorine or of Peroxide of Nitrogen in the place of Hydrogen. The peculiarity of these substances is, that they strongly resemble the originals in their physical, and often in their chemical properties. It might have been supposed that agents of such active chemical affinities as Chlorine and Oxide of Nitrogen would, by their mere presence in a body, produce a marked effect; yet it is not so in the case before us. The primitive type or constitution of the substance modified remains the same, even the crystalline form being often unaffected. It seems as if the body by which the Hydrogen had been displaced had stepped in quietly and taken up its position in the framework of the whole without disturbance. Many compounds of this kind are known; they are termed by chemists "substitution compounds." The law invariably observed is, that the substitution takes place *in equal atoms:* a single atom of Chlorine, for instance, displaces one of Hydrogen; two of Chlorine displace two of Hydrogen, and so on, until, in some cases, the whole of the latter element is separated.

In illustration of these remarks, take the following instances :—Acetic Acid contains Carbon, Hydrogen, and Oxygen; by the action of Chlorine the Hydrogen may be removed in the form of Hydrochloric Acid, and an equal number of atoms of Chlorine be substituted. In this way a new compound is formed, termed *Chloracetic Acid*, resembling in many important particulars the Acetic Acid itself. Notice particularly that the peculiar properties characteristic of Chlorine are completely masked in the substitution body, and no indication of its presence is obtained by the usual tests! A soluble *Chloride* gives with Nitrate of Silver a white precipitate of Chloride of Silver, unaffected by Acids, but the Chloracetic Acid does not; hence it is plain that the Chlorine exists in a peculiar and intimate state of combination different from what is usual.

The substance we have been previously considering, viz. Pyroxyline, affords another illustration of the Law of Substitution. Omitting, for the sake of simplicity, the *number* of atoms concerned in the change, the action of concentrated Nitric Acid upon ligneous fibre may be thus explained :—

$$\text{Cotton } or \begin{cases} \text{Carbon} \\ \text{Hydrogen} \\ \text{Hydrogen} \\ \text{Oxygen} \end{cases} + \quad \text{Nitric Acid}$$

equals

$$\text{Pyroxyline } or \begin{cases} \text{Carbon} \\ \text{Hydrogen} \\ \text{Peroxide Nitrogen} \\ \text{Oxygen} \end{cases} + \text{Water}$$

Or in symbols :—

$$CH_nO + NO_5 = C\,(H_{n-1}NO_4)\,O + HO$$

By a reference to the formula, it is seen that the fifth atom of Oxygen contained in the Nitric Acid takes one of

Hydrogen, and forms an atom of Water; the NO_4 then steps in, to fill the gap which the atom of Hydrogen has left. All this is done with so little disturbance that even the fibrous structure of the cotton remains as before.

b. *Chemical Composition of the varieties of Pyroxyline.* —Mr. Hadow has succeeded in establishing *four* different substitution compounds, which, as no distinctive nomenclature has been at present proposed, may be termed compounds *A, B, C,* and *D.*

Compound A is the most explosive Gun Cotton, and contains the largest amount of Peroxide of Nitrogen. It dissolves *only in Acetic Ether,* and is left on evaporation as a white powder. It is produced by the strongest Nitro-Sulphuric Acid which can be made.

Compounds B and C, either separate or in a state of mixture, form the soluble material employed by the Photographer. They both dissolve in Acetic Ether, and also in a mixture of Ether and Alcohol. The latter, viz. *C,* also dissolves in glacial Acetic Acid. They are produced by a Nitro-Sulphuric Acid slightly weaker than that used for *A,* and contain a smaller amount of Peroxide of Nitrogen.

Compound D resembles what has been termed *Xyloidine,* that is, the substance produced by acting with Nitric Acid upon Starch. It contains less Peroxide of Nitrogen than the others, and dissolves in Ether and Alcohol, and also in Acetic Acid. The ethereal solution leaves, on evaporation, an *opaque* film, which is highly combustible, but not explosive.

By bearing in mind the properties of these compounds, many of the anomalies complained of in the manufacture of Gun Cotton disappear. If the Nitro-Sulphuric Acid employed is too strong, the product will be insoluble in Ether; whilst if it is too weak, the fibres are gelatinized by the Acid and partly dissolved.

c. *Means adopted to procure a Nitro-Sulphuric Acid of the requisite strength for preparing Pyroxyline.*—This is a point of more difficulty than would at first appear. It is easy to determine an exact formula for the mixture, but not so easy to hit upon the proper proportions of the acids required to produce that formula; and a very slight departure from them altogether modifies the result. The main difficulty lies in *the uncertain strength of commercial Nitric Acid.* Oil of Vitriol is more to be depended upon, and has a tolerably uniform Sp. Gr. of 1·836 ;* but Nitric Acid is constantly liable to variation; hence it becomes necessary to make a preliminary determination of its real strength, which is done either by taking the specific gravity and referring to tables, or, better still, by a direct analysis. As each atom of Sulphuric Acid removes only a given quantity of water, it follows that the weaker the Nitric Acid, the larger the amount of Sulphuric which will be required to bring it up to the proper degree of concentration.

To avoid the trouble necessarily attendant upon these preliminary operations, many prefer to use, in place of Nitric Acid itself, one of the salts formed by the combination of Nitric Acid with an alkaline base. The composition of these salts, provided they are pure and nicely crystallized, can be depended on.

Nitrate of Potash, or *Saltpetre,* contains a single atom of Nitric Acid united with one of Potash. It is an *anhydrous* salt, that is, it has no water of crystallization. When strong Sulphuric Acid is poured upon Nitrate of Potash in a state of fine powder, in virtue of its superior chemical affinities it appropriates to itself the Alkali and liberates the Nitric Acid. If care be taken to add a sufficient excess of the Sulphuric Acid, a solution is obtained containing Sulphate of Potash dissolved in Sulphuric Acid, and free Nitric Acid. The presence of the Sulphate of Potash (or, more strictly speaking, of the *Bi*-Sulphate) does not in any way interfere with the result, and the effect is the same as if the mixed acids themselves had been used.

---

* The later experience of the writer induces him to believe, that the specific gravity of Oil of Vitriol cannot always be taken as an indication of its real strength; which is best ascertained by analysis.

The reaction may be thus represented :—

    Nitrate of Potash *plus* Sulphuric Acid in excess
= Bisulphate Potash *plus* Nitro-Sulphuric Acid.

CHEMISTRY OF THE SOLUTION OF PYROXYLINE IN ETHER AND ALCOHOL, OR "COLLODION."

The substitution compounds B and C, already alluded to as forming the Soluble Cotton of Photographers, are both abundantly soluble in Acetic Ether. This liquid however is not adapted for the purpose required, inasmuch as on evaporation it leaves the Pyroxyline in the form of a white powder, and not as a transparent layer.

The rectified Ether of commerce has been found to answer better than any other liquid as a solvent for Pyroxyline.

If the sp. gravity be about ·750, it contains invariably a small proportion of *Alcohol*, which appears to be necessary; the solution not taking place with absolutely pure Ether. The Pyroxyline, if properly prepared, begins almost immediately to gelatinize by the action of the Ether, and is soon completely dissolved. In this state it forms a slimy solution, which, when poured out on a glass plate, dries up into a horny transparent layer.

In preparing Collodion for Photographic purposes, we find that its physical properties are liable to considerable variation. Sometimes it appears very thin and fluid, flowing on the glass almost like water, whilst at others it is thick and glutinous. The causes of these differences will now engage our attention. They may be divided into two classes : first, those relating to the Pyroxyline; second, to the solvents employed.

    a. *Variation of Properties in different Samples of soluble Pyroxyline.*—The substitution compounds A, B, C, and D differ, as already shown, in the percentage amount of Peroxide of Nitrogen present, and the former are more explosive and insoluble than the latter. But it often happens in preparing Pyroxyline, that two portions of Nitro-Sulphuric Acid taken from the same bottle yield products which vary in properties, although they are necessarily the same in composition.

Taking *extremes* in illustration, we notice two principal modifications of soluble Pyroxyline.

The first, when treated with the mixture of Ether and Alcohol, sinks down to a gummy or gelatinous mass, which gradually dissolves on agitation. The solution is very fluid in proportion to the number of grains used, and when poured out spreads into a beautifully smooth and glassy surface, which is quite structureless, even when highly magnified. The film adheres tightly to the glass, and when the finger is drawn across it, separates in short fragments, and broken pieces.

The second variety produces a Collodion which is thick and glutinous, flowing over the glass in a slimy manner, and soon setting into numerous small waves and cellular spaces. The film lies loose upon the glass, is apt to contract on drying, and may be pushed off by the finger in the form of a connected skin.

This subject is not thoroughly understood, but it is known that the *temperature* of the Nitro-Sulphuric Acid at the time of immersing the Cotton influences the result. The soluble variety is produced by *hot* acids; the second, or glutinous, by the same acids employed cold, or only slightly warm. The best temperature appears to be from 130° to 155° Fahrenheit; if it rises much beyond that point, the acids act upon and dissolve the Cotton.

b. *The physical properties of Collodion affected by the proportions and purity of the Solvents.*—Pyroxyline of the varieties termed B and C dissolves freely in a mixture of Ether and Alcohol; but the characters of the resulting solution vary with the relative proportions of the two solvents.

When the Ether is in large excess, the film is inclined to be strong and tough, so that it can often be raised by

one corner and lifted completely off the plate without tearing. It is also very contractile, so that a portion of the
Collodion poured on the hand draws together and puckers
the skin as it dries. If spread upon a glass plate in the
usual way, the same property of contractility causes it to
retract and separate from the sides of the glass.

These properties, produced by Ether in large proportion,
disappear entirely on the addition of more Alcohol. The
transparent layer is now soft and easily torn, possessing
but little coherency. It adheres to the surface of the glass
more firmly, and exhibits no tendency to contract and separate from the sides.

From these remarks it will be gathered that an excess of
Ether, and a low temperature in preparing the Proxyline,
both favour the production of a contractile Collodion;
whilst on the other hand an abundance of Alcohol, and a
hot Nitro-Sulphuric Acid, tend to produce a short and
non-contractile Collodion.

The physical properties of Collodion are affected by another cause, viz. by the *strength* and purity of the solvents,
or, in other words, their freedom from dilution with water.
If a few drops of water be purposely added to a sample of
Collodion, the effect is seen to be to precipitate the Pyroxyline in flakes to the bottom of the bottle. There are
many substances known in chemistry which are soluble
in spirituous liquids, but behave in the same manner as
Pyroxyline in this respect.

The manner in which water gains entrance into the
Photographic Collodion is usually by the employment of
Alcohol or Spirit of Wine which has not been highly
rectified. In that case the Collodion is thicker, and flows
less readily than if the Alcohol were stronger. Sometimes
the texture of the film left upon evaporation is injured;
it is no longer homogeneous and transparent, but semi-
opaque, reticulated, or honeycombed, and so rotten that
a stream of water projected upon the plate washes it away.

These effects are to be attributed not to the Alcohol, but

to the water introduced with it; and the remedy will be
to procure a stronger spirit, or, if that cannot be done, to
increase the amount of Ether. Collodion prepared with a
large proportion of Ether, and water, but a small quantity
of Alcohol, is often very fluid and structureless at first,
adhering to the glass with some tenacity and having a short
texture; but it tends to become rotten when used to coat
many plates successively, the water on account of its
lesser volatility accumulating in injurious quantity in the
last portions.

THE COLORATION OF IODIZED COLLODION EXPLAINED.

Collodion iodized with the Iodides of Potassium, Ammonium, or Zinc, soon assumes a yellow tint, which in the
course of a few days or weeks, according to the temperature of the atmosphere, deepens to a full brown. This gradual coloration, due to a development of Iodine, is caused
partly by the Ether and partly by the Pyroxyline.

Ether may, with proper precautions, be preserved for a
long time in a pure state, but on exposure to the joint
action of air and light it undergoes a slow process of oxidation, attended with formation of Acetic Acid and a peculiar principle resembling in properties *ozone*, or Oxygen
in an allotropic and active condition. Iodide of Potassium
or Ammonium is decomposed by Ether in this state, Acetate of the Alkali, and Hydriodic Acid (HI), being first
produced. The ozonized substance then removes *Hydrogen*
from the latter compound, and liberates Iodine, which dissolves and tinges the liquid yellow.

A simple solution of an Alkaline Iodide in Alcohol and
Ether does not, however, become so quickly coloured as
Iodized Collodion ; and hence it is evident that the presence of the Pyroxyline produces an effect. It may be
shown that Alkaline Iodides slowly decompose Pyroxyline,
and that a portion of Peroxide of Nitrogen is set free:
this body, containing loosely combined oxygen, tends
powerfully to eliminate Iodine, as may be seen by adding

a few drops of the yellow commercial *Nitrous* acid to a solution of Iodide of Potassium.

The *stability* of the particular Iodide used in Iodizing Collodion, influences mainly the rate of coloration, though elevation of temperature and exposure to light are not without effect. Iodide of Ammonium is the least stable, and Iodide of Cadmium the most so; Iodide of Potassium being intermediate. Collodion iodized with *pure* Iodide of Cadmium usually remains nearly colourless to the last drop, if kept in a cool and dark place.

As the presence of free Iodine in Collodion affects its photographic properties, it may sometimes be necessary to remove it. This is done by inserting a strip of Silver foil; which decolorizes the liquid, by forming Iodide of Silver, soluble in the excess of Alkaline Iodide (p. 42). Metallic Cadmium, and metallic Zinc, have the same effect.

When Methylated Spirits are employed in the manufacture of Collodion, the Iodine first liberated is afterwards either partially or entirely reabsorbed, the liquid acquiring at the same time an acid reaction to test-paper.

SECTION II.

The Chemistry of the Nitrate Bath.

The solution of Nitrate of Silver in which the plate coated with iodized Collodion is dipped, to form the layer of Iodide of Silver, is known technically as *the Nitrate Bath.* The chemistry of Nitrate of Silver has been explained at page 13, but there are some points relating to the properties of its aqueous solution which require a further notice.

Solubility of Iodide of Silver in the Nitrate Bath.— Aqueous solution of Nitrate of Silver may be mentioned in the list of solvents of Iodide of Silver. The proportion dissolved is in all cases small, but it increases with the *strength* of the solution. If no attention were paid to this point, and the precaution of previously saturating the Nitrate Bath with Iodide of Silver neglected, the film would be dissolved when left too long in the liquid.

This solvent power of Nitrate of Silver on the Iodide is well shown by taking the excited Collodion plate out of the Bath, and allowing it to dry spontaneously. The layer of Nitrate on the surface, becoming concentrated by evaporation, eats away the film, so as to produce a transparent, spotted appearance.

In the solution of Iodide of Silver by Nitrate of Silver a *double salt* is formed, which corresponds in properties to the double Iodide of Potassium and Silver in being *decomposed* by the addition of water. Consequently, in order to saturate a Bath with Iodide of Silver it is only necessary to dissolve the total weight of Nitrate of Silver in a small bulk of water, and to add to it a few grains of an Iodide; perfect solution takes place, and on subsequent dilution with the full amount of water, the excess of Iodide of Silver is precipitated in the form of a milky deposit.

Acid condition of Nitrate of Silver.—A solution of *pure Nitrate of Silver* is neutral to blue litmus-paper, but that prepared from the commercial Nitrate has usually an acid reaction; the crystals having been imperfectly drained from the acid mother-liquor in which they were formed. Hence, in making a new Bath it is often advisable not only to saturate it with Iodide of Silver, but to neutralize the free Nitric acid it contains.

There is also a peculiar condition of Nitrate of Silver crystallized from a solution of the metal in Nitric Acid, which renders it quite unfit for photographic purposes (see p. 101). It is thought to depend upon the presence of an oxide of Nitrogen, possibly of Nitrous Acid, and the remedy is to dry the crystals very strongly, or, better still, to fuse them at a moderate heat: mere neutralization with Carbonate of Soda does not suffice.

In melting Nitrate of Silver great care should be taken not to raise the heat so high as to decompose the salt, or a basic Nitrite will be formed, which affects the properties of the solution (p. 13): fused Nitrate of Silver ought, when cold, to be quite white, and to dissolve perfectly in water without leaving any residue. The only objection to the employment of Nitrate of Silver in this form is the facility with which it may be adulterated with Nitrates of Potash and Soda, the presence of which would lessen the available strength of the Bath.

The Nitrate Bath, although perfectly neutral when first prepared, may become *acid* by continued use, if Collodion containing much *free Iodine* be constantly employed. In that case a portion of Nitric Acid is liberated, thus :—

Nitrate of Silver+Iodine
= Iodide of Silver +Nitric Acid+Oxygen.

When Collodion is iodized entirely with alkaline Iodides, it liberates Iodine by keeping; and hence the occasional use of Ammonia may be required to remove acidity from the Bath. But since the introduction of the Iodide of Cadmium, which preserves the Collodion nearly or quite colourless, the necessity for neutralizing Nitric Acid in the Bath has ceased.

Alkaline condition of the Bath.—By "alkalinity" of the Bath is meant a condition in which the blue tint is rapidly restored to reddened litmus-paper. This indicates that an Oxide of some kind is present in solution, which, by combining with the acid in the reddened paper, neutralizes it and removes the red colour.

If a small portion of caustic Potash or Ammonia be added to a strong solution of Nitrate of Silver, it produces a brown precipitate, which is Oxide of Silver.

Ammonia + Nitrate of Silver
= Oxide of Silver + Nitrate Ammonia.

The solution however, from which the precipitate has separated, is not left in a neutral state, but possesses a faint alkaline reaction. Oxide of Silver and Carbonate of Silver are also *abundantly* soluble in water containing Nitrate of Ammonia; which salt is continually accumulating

in the Bath when compounds of Ammonium are used for
iodizing.

An alkaline Bath is perhaps of all conditions the one
most fatal to success in photography.   It leads to that
universal darkening of the film on applying the developer
to which the name of "fogging" has been given.   Hence
care must be used in adding to the Bath substances which
tend to make it alkaline.

Collodion containing free Ammonia, often sold in the
shops, gradually does so.   The use of Potash, Carbonate
of Soda, Chalk, or Marble, to remove free Nitric Acid from
the Bath, has the same effect; and hence, when they are
employed, a trace of Acetic acid must afterwards be added.

The mode of testing a bath for alkalinity is as follows :—
a strip of porous blue litmus-paper is taken and held to
the mouth of a bottle of glacial Acetic acid until it becomes
reddened; it is then placed in the liquid to be examined
and left for ten minutes or a quarter of an hour.   If Oxide
of Silver be present in solution, the original blue colour of
the paper will slowly but gradually be restored.

*Occasional formation of Acetate of Silver in the Nitrate
Bath.*—In preparing a new Bath, if the crystals of Nitrate
of Silver are acid, it is usual to add an alkali in small
quantity.   This removes the Nitric Acid, but leaves the
solution faintly alkaline.   Acetic Acid is then dropped in,
which, by combining with the Oxide of Silver, forms
Acetate of Silver.

Acetate of Silver is not formed by the simple addition of
Acetic Acid to the Bath, because its production under such
circumstances would imply the liberation of Nitric Acid;
but if an alkali be present to neutralize the Nitric Acid,
then the double decomposition takes place, thus—

　　　　Acetate of Ammonia + Nitrate of Silver
　　= Acetate of Silver + Nitrate of Ammonia.

Acetate of Silver is a white flaky salt, sparingly soluble in water.   It dissolves in the Bath only in small
proportion, but yet sufficiently to affect the Photographic

properties of the film (see p. 111 and 117). The observance of the following simple rules will obviate its production in injurious quantity :—*First*, when it is required to remove free Nitric Acid from a bath *not containing Acetic Acid* a solution of Potash or Carbonate of Soda may be dropped in *freely;* but the liquid must be filtered before adding any Acetic Acid, otherwise the brown deposit of Oxide of Silver will be taken up by the Acetic Acid, and the Bath will be charged with Acetate of Silver. *Secondly*, in dealing with a Bath containing both Nitric and Acetic Acids, employ an alkali *much diluted* (Liquor Ammoniæ with 10 parts of Water), and add a single drop at a time, coating and trying a plate between each addition; the Nitric Acid will neutralize itself before the Acetic, and with care there will be no formation of Acetate of Silver in quantity.

Substances which decompose the Nitrate Bath.—Most of the common metals, having superior affinity for Oxygen, separate the Silver from a solution of the Nitrate; hence the Bath must be kept in glass, porcelain, or gutta-percha, and contact with Iron, Copper, Mercury, etc., must be avoided, or the liquid will be discoloured, and a black deposit of metallic Silver precipitated.

All developing agents, such as Gallic and Pyrogallic Acids, the Protosalts of Iron, etc., blacken the Nitrate Bath, and render it useless by reducing metallic Silver.

Chlorides, Iodides, and Bromides produce a deposit in the Bath; but the solution, although weakened, may again be used after passing through a filter.

Hyposulphites, Cyanides, and all fixing agents decompose Nitrate of Silver.

Organic matters, generally, reduce Nitrate of Silver, either with or without the aid of light. Grape Sugar, Albumen, Serum of Milk containing caseine, etc., blacken the Bath, even in the dark. Alcohol and Ether act more slowly, and produce no injurious effect unless the liquid is constantly exposed to light.

These facts indicate that the Nitrate Bath containing

volatile organic matters must be preserved in a dark place; also that it should be kept exclusively for sensitizing the Collodion plates, and not used in floating papers intended for the printing process.

Changes in the Nitrate Bath by use.—The solution of Nitrate of Silver employed in exciting the Collodion film gradually decreases in strength, but not so quickly as the Bath used in sensitizing papers for printing. If the amount of Nitrate be allowed to fall as low as twenty grains to the ounce of water, the decomposition will be imperfect, and the film will be pale and blue, even with a highly iodized Collodion.

A gradual accumulation of Ether and Alcohol also takes place in the Bath after long use, in consequence of which the developing solutions flow less readily upon the Collodionized plates, and oily stains are apt to be produced.

Diminished sensitiveness of the Iodide film is sometimes traced to impurities in the Bath, when it is very old, and has been much used. These are probably of an organic nature and may often be partially removed by agitation with kaolin, or animal charcoal. The latter however is objectionable, being usually contaminated with *Carbonate of Lime*, which makes the Bath alkaline; or (in the case of *purified* animal charcoal) with traces of Hydrochloric Acid, which liberate Nitric Acid in the Bath. Even the kaolin may as a preliminary precaution be washed with dilute Acetic Acid to remove Carbonate of Lime if any should be present.

SECTION III.

The Conditions which affect the Formation and Development of the Latent Image in the Collodion process.

It will be necessary to preface the observations contained in this Section by defining two terms which are frequently confounded with each other, but are in reality of distinct meaning. These terms are " Sensitiveness " and " Intensity."

By Sensitiveness is meant a facility of receiving impression from very feeble rays of light, or of receiving it quickly from brighter rays.

Intensity, on the other hand, relates to the appearance of the finished Photograph, independently of the time taken to produce it,—*to the degree of opacity of the image*, and the extent to which it obstructs transmitted light.

It will be seen as we proceed that the conditions necessary to obtain extreme sensitiveness of the Iodide film are different from, and often opposed to, those which give the maximum intensity of image.

CAUSES WHICH INFLUENCE THE SENSITIVENESS OF IODIDE OF SILVER ON COLLODION.

Some of the most important are as follows :—

a. *The presence of free Nitrate of Silver.*—When the sensitive film is removed from the Nitrate Bath, the Iodide of Silver is left in contact with excess of Nitrate of Silver. The presence of this compound is not *essential* to the action of the light, since, if it be removed by washing in distilled water, the image may still be impressed. In such a case however the effect is produced slowly, and a longer exposure in the Camera is required.

The sensitiveness of the Iodide film does not increase uniformly with the amount of the excess of Nitrate of Silver, as measured by the strength of the Bath. It is found that no advantage in this respect can be gained by using a proportion of Nitrate of Silver greater than 30 or 35 grains to the ounce of water, although solutions of three times this strength have been sometimes employed.

It has been asserted that a chemically pure Iodide of Silver, which is unaffected in colour by the direct action of light, is also incapable of receiving the invisible image in the Camera; and that the sensitiveness of a washed Collodion film is due to a minute quantity of Nitrate of Silver still remaining. Iodide of Silver in the state in which it is thrown down on diluting with water a strong solution of the salt known as the double Iodide of Potassium and Silver,—and which must, from the mode of its preparation, be free from Nitrate of Silver,—is quite insensitive; but this form of Iodide differs from the other in colour, and not only so, but is likely to contain an excess of Iodide of Potassium. The application of a solution of Nitrate of Silver to this compound at once renders it sensitive to light.

b. *Free acids in the Nitrate Bath.*—Strong oxidizing agents, such as Nitric Acid, greatly diminish the sensibility of the film, and hence the importance of removing the free acid often met with in commercial samples of the Nitrate of Silver. The effect of even a single drop of strong Nitric Acid in an eight-ounce Nitrate Bath will be appreciable; and when the proportion is increased to one drop per ounce, it will be difficult to obtain a rapid impression.

Acetic Acid has far less effect upon the sensitiveness than Nitric Acid, and being found useful during the development of the image is commonly employed; but when great rapidity is desired, it should be added cautiously, and in a proportion very much less than that in the solution known as the Aceto-Nitrate of Silver, which contains about one drop of the glacial acid to each grain of Nitrate of Silver.

c. *Addition of certain organic matters.*—It has long been remarked that the use of bodies like Albumen, Gelatine, Caseine, etc., which combine with Oxides of Silver, retard the action of light upon Iodide of Silver; and the recent observations of the Author enable him to confirm this statement. It is probable that one cause, amongst others, of the great sensibility of the Collodion film is due to the fact that Pyroxyline is a substance peculiarly indifferent to the Salts of Silver, exhibiting no tendency to reduce them to the metallic state; and it is proved by experiment that the addition of Grape Sugar, or of the resinous body, *Glycyrrhizine,* which resembles Albumen in causing a white precipitate in strong solution of Nitrate of Silver, renders necessary a longer exposure in the Camera. Alkaline Citrates have a still more marked effect, as also have Tartrates, Oxalates, etc.

d. *Impurities in the soluble Iodides.*—Commercial Iodide of Potassium often contains *Iodate* of Potash, which is found to have a retarding effect upon the action of light; also Carbonate of Potash, which, in Collodion, produces Iodoform,* and in the paper processes, where "Aceto-Nitrate" is used for sensitizing, forms Acetate of Silver. Iodoform has a marked influence in diminishing the sensitiveness of Iodide of Silver: Acetate of Silver may perhaps increase it a little by securing the absence of free Nitric Acid (p. 117). Iodide of Potassium prepared by the process in which Sulphuretted Hydrogen and Alcohol are used, and having a smell of Garlic, contains probably Xanthate of Potash, and is nearly useless for Photography.

Commercial Iodide of Cadmium is a purer salt than the Iodide of Potassium, and may be advantageously substituted for it; but it possesses the property of coagulating Albumen, and hence cannot be employed in conjunction with that substance.

e. *Presence of free Iodine.*—Both in the waxed paper and the Collodion processes, the solutions often contain a small quantity of free Iodine. This Iodine, in contact with the Nitrate of Silver of the Bath, produces a mixed Iodide and *Iodate* of Silver, and liberates Nitric Acid. It thus retards the sensitiveness of the film in proportion to the quantity of Iodine present. Collodion of a full yellow colour is perceptibly less sensitive than the same rendered colourless; and when enough Iodine has been liberated to give a red or brown tint, double the original exposure will probably be required.

If brown Collodion be much used, the Nitrate Bath may by degrees become sufficiently contaminated with free Nitric Acid to interfere with the sensitiveness of the film;

---

* See the Vocabulary, Part III., Art. Iodoform.

but if colourless or lemon-yellow tinted Collodion be employed, this evil need not be anticipated.

Certain substances may be added to coloured Collodion which possess the property of counteracting the retarding influence of the free Iodine, such, for instance, as the Oils of Cloves, Cinnamon, etc.; they probably act in virtue of their affinity for Oxygen, by preventing the formation of Iodate of Silver. In colourless Collodion they produce little or no effect, neither do they remove the insensitiveness of the film when dependent upon a too acid condition of the Nitrate Bath.

f. *Addition of Bromide or Chloride to Collodion.*—In the Daguerreotype a very exalted state of sensibility is obtained by exposing the silvered plate first to the vapour of Iodine, and afterwards to that of Bromine or Chlorine; but this rule does not apply to the Collodion process, which differs essentially in principle. Soluble Bromides added to Collodion lessen its sensibility to an appreciable extent, as also do Chlorides. This rule however may perhaps be liable to an exception when artificial light is used, which contains a greater proportion of the rays of small refrangibility, known to act more powerfully upon the Bromide than upon the Iodide of Silver (p. 66).

g. *Density of the sensitive film.*—When the proportion of soluble Iodide in the iodizing solution is too great, the film is very dense, and the Iodide of Silver is apt to burst out upon the surface, and fall away in loose flakes into the Bath. This condition, which is highly unfavourable to sensitiveness, is very common in Collodion, and constitutes what is termed " over-iodizing." The Iodide, in fact, is formed in such a case too much upon the surface, and consequently, when the fixing agent is applied, the image not being retained by the film, is washed off and lost.

On the other hand, the sensibility of the film is not lessened by reducing the amount of Iodide in Collodion to a minimum, if all the solutions are neutral; but the pale blue films formed by a dilute Collodion, and which almost

rival the Daguerreotype itself in delicacy, are nearly useless in practice; for if free Iodine or other bodies of a retarding nature are present in any quantity, either in the
Collodion or in the Bath, they almost destroy the action
of a weak light, producing a far more injurious effect than
if the film were more yellow and opaque.

h. *Impurities in Ether and Alcohol.*—Pure Ether should
be neutral to test-paper, but the commercial samples of
this article have usually either an acid or an alkaline reaction. The frequent occurrence of a peculiar oxidizing
principle in Ether has also been pointed out (p. 85). Each
of these three conditions is injurious to sensitiveness; the
first and last by liberating Iodine when alkaline Iodides
are used; and the second, by producing Iodoform under
the same circumstances. In this case the Collodion remains colourless, but gives inferior results.

The Author has also observed that Ether which has
been redistilled from the residues of Collodion may contain a volatile principle (probably a compound Ether?)
which produces a retarding effect upon the action of light.

Commercial Spirit of Wine is not always uniform in
composition, as sufficiently evidenced by the test of smell.
It may contain "fusel oil" or other volatile substances,
which become milky on dilution with water, and are be-
lieved to injure the quality of the spirit for Photographic
use.

i. *Relative proportions of Ether and Alcohol in Collodion.*—It was shown at p. 84 that the addition of Alcohol
to Collodion lessens the contractility of the film, and
renders it soft and gelatinous. This condition is favourable to the formation of the invisible image in the Camera,
the play of affinities being promoted by the loose manner
in which the particles of Iodide are held together. It is
therefore usual to add to Collodion as much Alcohol as
it will bear without becoming glutinous, or leaving the
glass; the exact quantity required varying with the strength
of the spirit or its freedom from dilution with water.

k. *Decomposition in the Collodion.*—Collodion iodized with the metallic Iodides generally, excepting the Iodide of Cadmium, becomes brown and loses sensitiveness in the course of a few days or weeks. If the free Iodine, the cause of the brown colour, be removed, the greater part, but not the whole, of the sensitiveness is regained. The experiments of the Author, and of others, have proved that a solution of Pyroxyline in contact with an unstable iodide, slowly undergoes decomposition, the result of which is that Iodine is set free, and an equivalent quantity of the base remains in union with certain organic elements of the Collodion.

Decomposition also gradually ensues when iodized Collodion is placed in contact with reducing agents, such as Proto-iodide of Iron, Gallic Acid, Grape Sugar, Glycyrrhizine, etc., so that these combinations do not retain a constant sensibility for any length of time. Even plain Collodion uniodized cannot be preserved many months without a small but perceptible amount of change.

l. *Decomposition in the Nitrate Bath.*—A Collodion Nitrate Bath which has been much used, often gives a less sensitive film than when newly made. It is known also that many organic substances which reduce Nitrate of Silver, if added to the Bath, produce a state which is favourable to sensitiveness whilst the decomposition is taking place, but is eventually unfavourable; hence the solution will be injured by adding either Gallic or Pyrogallic Acid, and by organic matters generally if exposed to light.

Recapitulation.—The conditions most favourable to extreme sensitiveness of the Iodide of Silver on Collodion may be condensed as follows:—perfect neutrality of the solutions employed; a soft, gelatinous state of the film; absence of Chlorides and other salts which precipitate Nitrate of Silver; an undecomposed Collodion, containing no organic matter of that kind which is precipitated by basic Acetate of Lead, and combines with oxides of Silver.

H

THE CONDITIONS WHICH AFFECT THE DEVELOPMENT OF THE LATENT IMAGE.

The general theory of the development of a latent image by means of a reducing agent, having been simply explained in the third Chapter, may now be more fully examined in its application to the Iodide of Silver on Collodion.

a. *The presence of free Nitrate of Silver essential to the development.*—This subject has already been mentioned (p. 36). A sensitive Collodion plate, carefully washed in distilled water, is still capable of receiving the radiant impression in the Camera, but it does not admit of development until it has been redipped in the Bath, or treated with a reducing agent to which Nitrate of Silver has been added: and if the proportion of free Nitrate of Silver on a Collodion film be too small, the image will be feeble or altogether imperfect in parts, with patches of green or blue, due to deficient reduction.

b. *Comparative strength of Reducing Agents.*—No increase of power in a developer will suffice to bring out a perfect image on an under-exposed plate, or upon a film containing too little Nitrate of Silver. But there is considerable difference in the the length of time which the various developers require to act. Gallic Acid is the most feeble, and Pyrogallic Acid the strongest, producing at least four times more effect than an equal weight of the crystallized Protosulphate of Iron, and twenty times more than the Protonitrate of Iron.

c. *The effect of free Acid upon the development.*—Acids tend to retard the reduction of the image as well as to diminish the sensibility of the film to light. Nitric Acid especially does so, from its powerful oxidizing and solvent properties. The effect of Nitric Acid is particularly seen when the film of Iodide of Silver is very blue and transparent, and the quantity of Nitrate of Silver retained upon its surface small. Under such circumstances the proper development of the image may be suspended, and spangles of metallic Silver separate. This indicates that the quantity of the acid should be diminished, or the strength of the Nitrate Bath and of the reducing agent be increased, as a counterpoise to the retarding action of acid upon the development.

Acetic Acid also moderates the rapidity of development, but it has not that tendency altogether to suspend it possessed by Nitric Acid. It is therefore usefully employed, to enable the operator to cover the plate evenly with liquid before the development commences, and to preserve the white parts of the impression from any accidental deposit of metallic Silver due to irregular action of the reducing agent.

On comparing the retarding effects of free acid upon the light's action, and upon the development, we see that the former is the most marked,—that a small quantity of Nitric Acid produces a more decided influence upon the impression of the image in the Camera than upon the bringing out of that image by means of a developer.

d. *Accelerating effect of certain organic matters.*—Organic bodies, like Albumen, Gelatine, Glycyrrhizine, etc., which combine chemically with oxides of Silver, and were shown in the last Section to lessen the sensitiveness of the Iodide film,—facilitate the development of the image, producing often a dense deposit of a brown or black colour by transmitted light.

In the same way, viz. by a retention of organic matter, may partly be explained the fact, that the image developed by Pyrogallic Acid, although proved by the application of tests to contain no more than an equal quantity of Silver, possesses greater opacity by transmitted light, than that resulting from the use of protosalts of Iron: and in the case of the Collodion itself the same rule applies— if it be pure, it is liable to give a less vigorous impression than when by long keeping a partial decomposition has taken place, and products have been formed which combine with reduced Oxide of Silver more easily than the unaltered Pyroxyline.

e. *Molecular conditions affecting Intensity.*—The physical structure of the Collodion film is thought to exert an influence upon the mode in which the reduced Silver is thrown down during the development. A short and almost powdery state, such as Collodion iodized with the alkaline iodides acquires by keeping, is considered favourable, and a glutinous, coherent structure unfavourable, to density. This is certainly the case when the film is allowed to dry before development, as in the process with desiccated Collodion and, to some extent, in the Oxymel preservative process.

The mode of conducting the development also affects the density; a rapid action tending to produce an image of which the particles are finely divided and offer a considerable resistance to the passage of light, whilst a slow and prolonged development often leaves a metallic and almost crystalline deposit, comparatively translucent and feeble.

The writer has observed, that with certain samples of Collodion the image is much enfeebled by keeping the plate for a considerable time,—a quarter of an hour or longer,—after sensitizing, but before development. This effect is not the result of the Nitrate of Silver having partially drained away, since a second dip in the Nitrate Bath immediately before applying the Pyrogallic Acid, does not remedy it. An alteration of molecular structure may therefore be the correct explanation, and if so, a contractile Collodion would suffer more than one possessing less coherency.

The actinic power of the light at the time of taking the picture, influences the appearance of the developed image; the most vigorous impressions being produced by a strong light acting for a short time. On a dull dark day, or in copying badly lighted interiors, the photograph will often lack bloom and richness, and be blue and inky by transmitted light.

f. Development of images upon Bromide and Chloride of Silver.—Of the three principal Salts of Silver, the Iodide is the most sensitive to light, but the Bromide and Chloride, under some conditions, are more easily developed and give a darker image. In the Collodion process the difference is principally seen when organic bodies, like Grape Sugar, Glycyrrhizine, etc., are introduced in order to increase the intensity; a far more decided effect being produced by adding both Glycyrrhizine and a portion of Bromide or Chloride, than by using the Glycyrrhizine alone.[*]

g. The intensity of the image affected by the length of exposure.—This point has been briefly alluded to in the third Chapter. If the exposure in the Camera be prolonged beyond the proper time the development takes place rapidly but without any intensity, the picture being pale and translucent. The effects produced by over-action of the light are particularly seen when the Nitrate Bath contains Nitrite of Silver, or Acetate of Silver; the image being frequently in such a case dark by reflected light, and red by transmitted light,—more nearly resembling in fact a photographic print, developed on paper prepared with Chloride of Silver. When Collodion plates are coated with honey without previously removing the free Nitrate of Silver, a slow reducing action is set up, which may give rise to the characteristic appearance above referred to, after development. Other organic substances, such as biliary matter, etc., will act in the same way.

h. Certain conditions of the Bath which affect development.—Attention may be called to a peculiar state of the Nitrate Bath, in which the Collodion image developes unusually slowly, and has a dull grey metallic appearance, with an absence of intensity in the parts most acted on by the light. This condition, which occurs only when using a newly mixed solution, is thought by the Author to depend

---

[*] See the Author's Paper on the chemical composition of the photographic image, in the eighth Chapter.

upon the presence of an Oxide of Nitrogen retained by the Nitrate of Silver. It is removed partially by neutralizing the Bath with an alkali, more perfectly so by adding an excess of alkali followed by Acetic Acid; but most completely by carefully *fusing* the Nitrate of Silver before dissolving it.

Commercial Nitrate of Silver has sometimes a fragrant smell, similar to that produced by pouring strong Nitric Acid upon Alcohol. When such is the case, it contains organic matter, and produces a Bath which yields red and misty pictures.

Nitrate of Silver which has been sufficiently strongly fused to decompose the Salt, and produce a portion of the basic Nitrite of Silver exhibits great peculiarity of development, the image coming out instantaneously and with great force. This condition is exactly the reverse of that produced by the presence of acids, in which the development is slow and gradual.

In summing up the different conditions of the Nitrate Bath which affect the development of the image, as many as *four* might be mentioned, each of which gives a more rapid reduction than the one which precedes it. These are—the acid Nitrate Bath, the neutral Bath, the Bath of strongly fused Nitrate of Silver, and the Bath containing *Ammoniacal* Nitrate of Silver, which is quite unmanageable, and produces an instantaneous and universal blackening of the film on the application of the developer.

Greater intensity of image is commonly obtained in a Nitrate Bath which has been a long time in use, than in a newly mixed solution: this may be due to minute quantities of organic matter dissolved out of the Collodion film, which, having an affinity for Oxygen, partially reduce the Nitrate of Silver; and also to the accumulation of Alcohol and Ether in an old Bath producing a short and friable structure of the film.

i. *Effect of Temperature on Development.*—Reduction of the oxides of noble metals proceeds more rapidly in proportion as the temperature rises. In cold weather it will be found that the development of the image is slower than usual, and that greater strength of the reducing agent and more free Nitrate of Silver is required to produce the effect.

On the other hand, if the heat of the atmosphere be excessive, the tendency to rapid reduction will be greatly increased, the solutions decomposing each other almost immediately on mixing. In this case the remedy will be to use Acetic Acid *freely* both in the Bath and in the developer, at the same time lessening the quantity of Pyrogallic Acid, and omitting the Nitrate of Silver which is sometimes added towards the end of the development.

Also in the case of films which are to be kept for a long time in a sensitive condition by means of honey, etc., the modifying influence of temperature must be observed, and the quantity of free Nitrate of Silver left upon the film be reduced to a minimum if the thermometer stands higher than usual.

SECTION IV.

On certain Irregularities in the Developing Process.

The characteristics of the proper development of a latent image are—that the action of the reducing agent should cause a blackening of the Iodide in the parts touched by light, but produce no effect upon those which have remained in shadow.

In operating both on Collodion and paper however there is a liability to failure in this respect; the film beginning, after the application of the developer, to change in colour to a greater or less extent over the whole surface.

There are two main causes which produce this state of things :—the first being due to an irregularity in the action of the light; the second to a faulty condition of the chemicals employed.

If from a defect in the construction of the instrument,

or from other causes which will be pointed out more particularly in the Second Part of this work, diffused white
light gains entrance into the Camera, it produces indistinctness of the image by affecting the Iodide more or less universally.

The luminous image of the Camera not being perfectly
pure, mere *over-exposure* of the sensitive plate will usually
have the same effect. In such a case, when the developer
is poured on, a faint image first appears, and is followed by
a general cloudiness.

The clearness of the developed Collodion picture is much
influenced by the condition of all the solutions employed,
but particularly so by that of the Nitrate Bath. If this
liquid be in the state termed alkaline (p. 88), it will be impossible to obtain a good picture; and even when neutral,
care and avoidance of all disturbing causes will be required to prevent a deposition of Silver upon the shadows
of the image: especially so when Nitrite of Silver or Acetate of Silver are present, both of these salts being more
easily reduced than the Nitrate of Silver.

The use of *Acid* is the principal resource in obviating
cloudiness of the image. Acids lessen the facility of reduction of the Salts of Silver by developing agents (p. 98), and
hence when they are present the metal is deposited more
slowly, and only on the parts where the action of the light
has so modified the particles of Iodide as to favour the
decomposition: whereas if acids be absent or present in
insufficient quantity, the equilibrium of the mixture of
Nitrate of Silver and reducing agent which constitutes
the developer is so unstable, that any rough point or sharp
edge is likely to become a centre from which the chemical
action, once started, radiates to all parts of the plate.

Various acids have been employed, such as Acetic acid,
Citric acid, Tartaric acid, etc. Nitric acid is the most
effectual of all, but is seldom used, because, although the
image can often be developed with great clearness when
the Bath contains a small quantity of Nitric acid, yet such

a condition is not favourable to *intensity;* on the other hand, films which are prone to irregular reduction; such as those prepared in a chemically neutral bath or a bath containing Acetate or Nitrite of Silver, are likely to give the greatest vigour of impression. Hence, when this quality is desired, the use of Nitric Acid will be adopted cautiously.

The state of the Collodion must be attended to as well as that of the bath; it should be either acid or neutral, not alkaline. Colourless Collodion may be used successfully as a rule, but sometimes a little free Iodine is advantageously added. Care should be taken in introducing organic substances, many of which dissolve out into the bath, and spoil it for giving clear pictures. Glycyrrhizine, however, which is recommended to produce intensity of Negatives, has no effect of that kind, and may be employed with safety.

The condition of the developing agent is a point of importance in producing clear and distinct pictures. The Acetic acid, which is advised in the formulæ, cannot be omitted or even lessened in quantity without danger. This is particularly the case in hot weather or under any other condition which favours reduction, such as neutrality of the bath, etc.; at all times, in fact, when the solutions of Pyrogallic acid and Nitrate of Silver decompose each other with unusual rapidity.

In addition to the points now mentioned, viz. the state of the Bath, of the Collodion, and of the developer, the reader should also study the remarks made in the Third Section of Chapter III. on the effect of *surface conditions* in modifying the deposition of vapour and of metallic Silver: he will then in all probability experience but little difficulty in dealing with those numerous irregularities in the action of the developing fluid, which often prove the greatest hindrance to the successful practice of the Collodion process.

CHAPTER VII.

ON POSITIVE AND NEGATIVE COLLODION PHOTO-
GRAPHS.

THE terms "Positive" and "Negative" occur so frequently in all works upon the subject of Photography, that it will be impossible for the student to make progress without thoroughly understanding their meaning.

A Positive may be defined to be a Photograph which gives a natural representation of an object, as it appears to the eye.

A Negative Photograph, on the other hand, has the lights and shadows reversed, so that the appearance of the object is changed or negatived.

In Photographs taken upon *Chloride of Silver*, either in the Camera or by superposition, the effect must necessarily be Negative; the Chloride being *darkened by luminous rays*, the lights are represented by shadows.

The following simple diagrams will make this obvious.

Fig. 1 is an opaque image drawn upon a transparent

Fig. 1.　　　　　Fig. 2.　　　　　Fig. 3.

ground; fig. 2 represents the effect produced by placing it
in contact with a layer of sensitive Chloride and exposing
to light; and fig. 3 is the result of copying this negative
again on Chloride of Silver.

Fig. 3 therefore is a Positive copy of Fig. 1, obtained by
means of a Negative.   By the first operation the tints are
reversed; by the second, being reversed again, they are
made to correspond to the original. The possession of a
Negative therefore enables us to obtain Positive copies of
the object, indefinite in number and all precisely similar in
appearance. This capability of multiplying impressions is
of the utmost importance, and has rendered the production
of good Negative Photographs of greater consequence
than any other branch of the Art.

The same Photograph may often be made to show either
as a Positive or as a Negative. For instance, supposing a
piece of silver-leaf to be cut into the shape of a cross and
pasted on a square of glass, the appearance presented by
it would vary under different circumstances.

Fig. 1 represents it placed on a layer of black velvet;

Fig. 1.   Fig. 2.

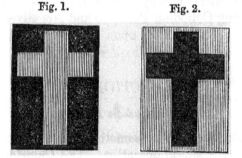

fig. 2 as held up to the light. If we term it Positive in
the first case, i. e. by reflected light, then it is Negative
in the second, that is, by transmitted light. The explanation is obvious.

Therefore to carry our original definition of Positives
and Negatives a little further, we may say, that the former
are usually viewed by reflected, and the latter by transmitted, light.

All Photographs however cannot be made to represent both Positives and Negatives. In order to possess this capability, it is necessary that a part of the image should be transparent, and the other opaque *but with a bright surface.* These conditions are fulfilled when the Iodide of Silver upon Collodion is employed, in conjunction with a developing agent.

Every Collodion picture is to a certain extent both Negative and Positive, and hence the processes for obtaining both varieties of Photographs are substantially the same. Although however the general characters of a Positive and a Negative are similar, there are some points of difference. A surface which appears perfectly opaque when looked down upon, becomes somewhat translucent on being held up to the light; hence, to give the same effect, the deposit of metal in a Negative must be proportionally thicker than in a positive; otherwise the minor details of the image will be invisible, from not obstructing the light sufficiently.

With these preliminary remarks, we are prepared to investigate more closely the *rationale* of the processes for obtaining Collodion Positives and Negatives. All that refers to paper Positives upon Chloride of Silver will be treated in a subsequent Chapter.

SECTION I.

On Collodion Positives.

Collodion Positives are sometimes termed *direct*, because obtained by a single operation. The Chloride of Silver, *acted upon by light alone*, is not adapted to yield direct Positives, the reduced surface being dark and incapable of representing the lights of a picture. Hence a developing agent is necessarily employed, and the Iodide of Silver substituted for the Chloride, as being a more sensitive preparation. Collodion Positives are closely allied in their nature to Daguerreotypes. The difference between the two consists principally in the surface used· to sustain the sensitive layer, and the nature of the substance by which the invisible image is developed.

In a Collodion Positive the lights are formed by a bright surface of reduced Silver, and the shadows by a black background showing through the transparent portions of the plate.

Two main points are to be attended to in the production of these Photographs.

First, to obtain an image distinct in every part, *but of comparatively small intensity*.—If the deposit of reduced metal be too thick, the dark background is not seen to a sufficient extent, and the picture in consequence is deficient in shadow.

Secondly, to *whiten* the surface of the reduced metal as much as possible, in order to produce a sufficient contrast of light and shade. Iodide of Silver developed in the usual way presents a dull yellow appearance which is sombre and unpleasing.

The Collodion and Nitrate Bath for Positives.—Good Positives may be obtained by diluting down a sample of Collodion with Ether and Alcohol until it gives a pale bluish film in the Bath. The proportion of Iodide of Silver being in that case small, the action of the high lights is less violent, and the shadows are allowed more time to impress themselves. The dilution lessens the amount of Pyroxyline in the Collodion at the same time with the Iodide, which is an advantage, the slight and transparent films always giving more sharpness and definition in the picture.

The employment of a very thin film for Positives is not however always a successful process. The particles of the Iodide of Silver being closely in contact with the glass, unusual care is required in cleaning the plates in order to avoid stains; and the amount of free Nitrate of Silver retained upon the surface of the film being small, circular patches of imperfect development are liable to occur, unless the reducing agent be scattered evenly and perfectly over the surface. Also if free Iodine or organic substances which have a retarding effect on the action of light are present to a considerable extent, the Collodion will not work well with a small proportion of Iodide. The Author found in experimenting on this subject that with perfectly pure Collodion and a *neutral* Bath most vigorous impressions were produced when the density of the film had been so far reduced by dilution that scarcely anything could be seen upon the glass; but with Collodion strongly tinted with Iodine, or with a Bath containing Nitric Acid, it was necessary to stop the dilution at a certain point or the film became absolutely insensitive to feeble radiations of light, and the shadows could not be brought out by any amount of exposure. In this case, by adding more Iodide a better effect was obtained.

A thicker Collodion may be used for positives if a little free Iodine be added, for the purpose of diminishing intensity and keeping the shadows clear during the development. This process is easier to practise than the last, but does not always give the same perfect definition.

No organic substance of the class to which Glycyrrhizine and Sugars belong should be added to Collodion which is to be used for Positives. By so doing the image would be rendered intense, and the high lights liable to solarization, *id est*, a dark appearance by reflected light.

The Nitrate Bath.—If the materials are pure, the *Nitrate Bath* may advantageously be diluted down at the same time with the Collodion, when Positives are to be taken; but the employment of a very weak Nitrate Bath (such as one of 20 grains to the ounce), although highly useful in obviating excess of development, has some disadvantages; it becomes necessary to exclude free Nitric Acid, and to avoid the employment of a Collodion too highly tinted with Iodine. On the other hand, with a strong Nitrate Bath, and a tolerably dense film of Iodide of Silver, a better result is often secured by the use of

Nitric Acid. The sensitiveness of the plates is impaired, but at the same time the intensity is diminished, and the picture shows well upon the surface of the glass.

A new Bath is better for taking Positives than one which has been a long time in use. The latter often causes *haziness* and irregular markings on the film during the action of the developer. This is due partly to the accumulation of Alcohol and Ether in the Bath, which causes the solution of Sulphate of Iron to flow in an oily manner; and partly to a reduction of the Nitrate of Silver by organic matter.

The presence of *Acetate of Silver* is objectionable in a Positive Nitrate Bath as producing solarization and intensity of image; hence those precautions which obviate its formation must be adopted (p. 89).

If fused Nitrate of Silver be used for the Positive Nitrate Bath, it is very important that the fusion should not be carried too far, or the solution would contain a basic Nitrite of Silver, and yield an intense, solarized, and misty image.

The Developers for Collodion Positives.—Pyrogallic Acid when used with Acetic Acid, as is usual for negative pictures, produces a surface which is dull and yellow. This may be obviated by substituting Nitric Acid *in small quantity* for the Acetic. The surface produced by Pyrogallic Acid with Nitric Acid is lustreless, but very white, if the solution be used of the proper strength. On attempting to increase the amount of Nitric Acid the deposit becomes metallic, and the half-tones of the picture are injured; Pyrogallic Acid, although an active developer, does not allow of the addition of mineral acid to the same extent as the Salts of Iron. It requires also, when combined with Nitric Acid, a fair proportion of Nitrate of Silver on the film, or the development will be imperfect in parts of the plate.

Sulphate of Iron.—The Protosalts of Iron were first employed in Photography by Mr. Hunt. The Sulphate is a most energetic developer, and often brings out a picture when others would fail. To produce by means of it a dead white tint with absence of metallic lustre, it may be used in conjunction with Acetic Acid, and in a somewhat concentrated condition, so as to develope the picture quickly.

The addition of *Nitric Acid* to Sulphate of Iron modifies the development, making it more slow and gradual, and producing a bright sparkling surface of reduced Silver. Too much of this acid however must not be used, or the action will be irregular. The Nitrate Bath also must be tolerably concentrated, in order to compensate for the retarding effect of Nitric Acid upon the development. The blue and transparent films of Iodide of Silver, formed in a very dilute Nitrate Bath, are not well adapted for Positives to be developed in this way. They are injured by the acid, and the development of the image becomes imperfect.

Protonitrate of Iron.—This salt, first used by Dr. Diamond, is remarkable as giving a surface of brilliant metallic lustre without any addition of free acid. Theoretically, it may be considered as closely corresponding to the Sulphate of Iron with Nitric Acid added. There are however slight practical differences between them, which are perhaps in favour of the Protonitrate.

The reducing powers of *Protoxide* of Iron appear to be in inverse ratio to the strength of the acid with which it is associated in its salts; hence the *Nitrate* is by far the most feeble developer of the Protosalts of Iron.

The rules already given for the use of Sulphate of Iron acidified with Nitric Acid, apply also to the Nitrate of Iron; the proportion of free Nitrate of Silver must be large, and the film of Iodide of Silver not too transparent.

In developing direct Positives either by Pyrogallic Acid or the Salts of Iron, the colour of the image will be found liable to some variation; the character of the light, whether bright or feeble, and the length of exposure in the Camera, affecting the result.

A Process for whitening the Positive Image by means of Bichloride of Mercury.—In place of brightening the Positive image by modifying the developer, it was proposed some time since by Mr. Archer to effect the same object by the use of the salt known as *Corrosive Sublimate*, or Bichloride of Mercury.

The image is first developed in the usual way, fixed, and washed. It is then treated with the solution of Bichloride, the effect of which is to produce almost immediately an interesting series of changes in colour. The surface first *darkens considerably*, until it becomes of an ash-grey, approaching to black; shortly it begins to get lighter, and assumes a *pure white* tint, or a white slightly inclining to blue. It is then seen, on examination, that the whole substance of the deposit is entirely converted into this white powder.

The *rationale* of the reaction of Bichloride of Mercury appears to be, that the Chlorine of the mercurial salt divides itself between the Mercury and the Silver, a portion of it passing to the latter metal and converting it into a Protochloride. The white powder is therefore probably a compound salt, as is further evidenced by the effects produced on treating it with various reagents.

SECTION II.

On Collodion Negatives.

As in the case of a direct Positive we require an image which is *feeble* though distinct, so, on the other hand, for a negative, it is necessary to obtain one of considerable intensity. In the Chapter immediately following the present, it will be shown that in using glass Negatives to produce Positive copies upon Chloride of Silver paper, a good result cannot be secured unless the Negative is sufficiently dark to obstruct light strongly.

The Collodion and Nitrate Bath for Negatives.—A Collodion containing a very small portion of Iodide and yielding a blue transparent film in the Bath is not well adapted for taking Negatives. Pale opalescent films often give too

I

little intensity in the high lights, and, unless the Nitrate Bath be acid, do not admit of being exposed in the Camera for the proper length of time without cloudiness and indistinctness of image being produced under the action of the developer. The effect known as " solarization of negatives," *i. e.* a red and translucent appearance of the highest lights, is also more liable to occur when operating with a very pale film. On the other hand, if the layer of Iodide be too yellow and creamy, the half-tones of the image will often be imperfectly developed, so that a middle point between these extremes is the best.

A pure and newly prepared Collodion, although highly sensitive to light, does not always give, with one application of the developer, a sufficiently vigorous image to serve as a negative matrix; and this particularly in the most brightly illuminated parts, such as the sky in a landscape photograph, or the white borders of an engraving. But on keeping the Collodion for some weeks or months it becomes yellow, if iodized with the alkaline iodides, and a decomposition takes place in it, as before shown (p. 97), which lessens the rapidity of action, but adds to the intensity of the negative.

Grape Sugar may be employed for the purpose of giving intensity to newly mixed Collodion: also Glycyrrhizine, which is a resinous body extracted from the root of Liquorice; but as both substances have an effect in lessening the sensitiveness and keeping qualities of the fluid, they should be used cautiously. In taking portraits in the open air, on bright days, and with a Bath which has been mixed for a considerable time, it will rarely be found that the intensity will be deficient; and especially so if the developer be applied a second time to the film with a few drops of solution of Nitrate of Silver added. In landscape Photography however, or in copying engravings, where extreme sensitiveness is not an object, the Glycyrrhizine may sometimes be added with advantage in order to obtain perfect opacity of the blacks.

When the use of this substance is resorted to, the mode of iodizing the Collodion appears to be of importance, the increase of intensity being greater with the Iodide of Cadmium than with the Iodides of the Alkalies; the latter probably exercising a decomposing action. An addition of a Bromide or a Chloride to the Collodion in small quantity has also a marked effect in adding to the intensity when Glycyrrhizine is used with alkaline Iodides (p. 101).

Substances which produce intensity of the Collodion image have often, if added in too large quantity, a tendency to lower the half-tone, and prevent the darker parts of the picture from being sufficiently brought out. The print from the Negative is then pale and white, or " chalky " as it is termed, in the high lights. Collodion in this condition is often preferred by the beginner, from the facility with which the Negatives are obtained, but it does not give the finest results. An excess of Glycyrrhizine in Collodion has also the effect of interfering with the precipitation of the Iodide of Silver, producing a blue and smoky film which is nearly useless for Negatives.

A judicious employment of free Iodine in Collodion which has been previously intensified with Glycyrrhizine, has a remarkable effect in improving the gradation of tone. The excessive opacity of the high lights is diminished, and hence the operator is enabled by a longer exposure of the sensitive plate to bring out the shadows and minor details of the image with great distinctness. Collodion prepared in this manner is too slow to be used for portraits, excepting in a strong light, but often gives an image with great roundness and stereoscopic effect.

The Iodide and the liquorice sugar employed conjointly, tend also to preserve the clearness of the plates under the influence of the developer, and to give sharpness to the lines and dots of engravings, etc., which, with a new and sensitive Collodion, are often imperfectly rendered. These advantages will be appreciated by the operator who has failed from working with a too feeble Collodion; but it must be borne in mind, that all substances acting as intensifiers have a bad effect when the state of the film is not such as to call for their employment.

The Proto-iodide of Iron has been recommended as an addition to Negative Collodion. In the Nitrate Bath it forms, in addition to Iodide of Silver, Protonitrate of Iron, an unstable substance and a developer. The use of Iodide of Iron gives great sensibility, but it is difficult to preserve it pure and unchanged. It also decomposes the Collodion in the course of a few hours, becoming itself peroxidized, and producing an insensitive condition of film. In addition to this, the negatives taken by the aid of Iodide of Iron are commonly of an inferior kind, the reduction being too marked in the high lights; so that its employment is of doubtful utility.

The Nitrate Bath.—This should be prepared from Nitrate of Silver which has been melted at a moderate heat (see pp. 13 and 101). If this point be neglected, the best Collodion will sometimes fail in producing an intense negative.

Acetic Acid must be added in minute quantity, to preserve the solution from a too ready reduction by the Alcohol and Ether of the Collodion. Also, unless the Nitrate of Silver be quite pure and free from organic matter (p. 104), clear pictures will not be obtained without the use of Acid.

Acetate of Silver has often been advised as an addition to the Negative Nitrate Bath. It is produced by dropping into the solution an alkali, such as Ammonia, followed by Acetic Acid in excess. The Negatives are rendered blacker and more vigorous by this proceeding, but especially so when the Bath is contaminated with Nitric Acid; which neutralizes itself at the expense of the Acetate of Silver, thus:—

Acetate of Silver + Nitric Acid
= Nitrate of Silver + Acetic Acid.

As a rule, it will be better to avoid adding Acetate of Silver to the Bath, since with pure melted Nitrate of Silver no Nitric Acid can be present, and perfect intensity is easily obtained. When the Bath is saturated with Acetate of Silver, it is in a more reducible state, and hence unless the glass plates are very perfectly cleaned, black lines and markings, the results of irregular action, will be produced on the application of the developer to the film (p. 104). *Solarization,* or reddening by over-exposure, is also promoted by the presence of Acetate of Silver.

Developing solutions for Negatives.—The Proto-salts of Iron are not usually employed in developing Negative impressions. They are liable to yield a violet-coloured image, which cannot easily be rendered more intense by continuing the action.

Gallic Acid is too feeble for developing Collodion pictures. Pyrogallic Acid is much superior, and may be used of any strength, according to the effect desired. When the light is bad, the temperature low, and the Negative developes slowly and appears blue and inky by transmitted light, the proportion of the reducing agent should be increased. But with an intense Collodion, on a clear summer's day, the finest gradation is obtained with a weak solution, which does not begin to act until the plate has been evenly covered. A strong developer might in such a case produce too much opacity in the highest lights, and would probably occasion stains of irregular reduction.

Modes of strengthening a finished impression which is too feeble to be used as a Negative.—The ordinary plan of pushing the development cannot be applied *with advantage* after the picture has been washed and dried. In that case, if it is found to be too feeble to print well, its intensity may be increased by one of the following methods.—

It must be premised however, that the same degree of excellence is not to be expected in a Negative Photograph which has been improperly developed in the first instance, and more especially if the exposure to light was too short.

Any "instantaneous Positive" may be rendered sufficiently intense for a Negative, but in that case *the shadows* are almost invariably imperfect.

1. *Treatment of the image with Sulphuretted Hydrogen or Hydrosulphate of Ammonia.*—The object is to convert the metallic Silver into *Sulphuret of Silver,* and if this could be done it would be of service. The mere application of an Alkaline Sulphuret has however but little effect upon the image, excepting to darken its surface and destroy the Positive appearance by reflected light; the structure of the metallic deposit being too dense to admit of the Sulphur reaching its interior.

Professor Donny ('Photographic Journal,' vol. i.) proposes to obviate this by first converting the image into the white Salt of Mercury and Silver by the application of Bichloride of Mercury, and afterwards treating it with solution of Sulphuretted Hydrogen or Hydrosulphate of Ammonia. Negatives produced in this way are of a brown-yellow colour by transmitted light, and opaque to chemical rays to an extent which would not, *à priori,* have been anticipated.

2. *MM. Barreswil and Davanne's process.*—The image is converted into Iodide of Silver by treating it with a saturated solution of Iodine in water. It is then washed—to remove the excess of Iodine,—exposed to the light, and a portion of the ordinary developing solution, mixed with Nitrate of Silver, poured over it. The changes which ensue are precisely the same as those already described; the whole object of the process being to bring the metallic surface back again into the condition of Iodide of Silver modified by light, that the developing action may be commenced afresh, and more Silver deposited from the Nitrate in the usual way.

3. *The process with Bichloride of Mercury and Ammonia.*—The image is first converted into the usual white double Salt of Mercury and Silver by the application of a solution of the Corrosive Sublimate. It is then treated

with Ammonia, the effect of which is to *blacken* it intensely. Probably the alkali acts by converting Chloride of Mercury into the black Oxide of Mercury. In place of Ammonia, a dilute solution of Hyposulphite of Soda or Cyanide of Potassium may be used, with very similar results.

CHAPTER VIII.

ON THE THEORY OF POSITIVE PRINTING.

THE subject of Collodion Negatives having been explained in the previous Chapter, we proceed to show how they may be made to yield an indefinite number of copies with the lights and shadows correct as in nature.

Such copies are termed "Positives," or sometimes "Positive prints," to distinguish them from direct Positives upon Collodion.

There are two distinct modes of obtaining photographic prints;—first by development, or, as it is termed, *by the Negative process*, in which a layer of Iodide or Chloride of Silver is employed, and the invisible image developed by Gallic Acid; and second, by the direct action of light upon a surface of Chloride of Silver, no developer being used. These processes, involving chemical changes of great delicacy, require a careful explanation.

The action of light upon Chloride of Silver was described in Chapter II. It was shown that a gradual process of darkening took place, the compound being reduced to the condition of a coloured *subsalt*; also, that the rapidity and perfection of the change were increased by the presence of excess of Nitrate of Silver, and of organic matters, such as Gelatine, Albumen, etc.

We have now to suppose that a sensitive paper has been prepared in this way, and that a Negative having been laid in contact with it, the combination has been ex-posed to the agency of light for a sufficient length of time. Upon removing the glass, a Positive representation of the object will be found below, of great beauty and detail. Now if this image were in its nature fixed and permanent, or if there were means of making it so, without injury to the tint, the production of Paper Positives would certainly be a simple department of the Photographic Art; for it will be found that with almost any Negative, and with sen-sitive paper however prepared, the picture will look tole-rably well on its first removal from the printing-frame. Immersion in the bath of Hyposulphite of Soda however, which is essentially necessary in order to fix the picture, produces an unfavourable effect upon the tint; decom-posing the violet-coloured Subchloride of Silver, and leav-ing behind a red substance which appears to be united to the fibre of the paper, and, when tested, reacts in the man-ner of a Suboxide of Silver.

Other chemical operations are therefore required to re-move the objectionable red colour of the print, and hence the consideration of the subject is naturally divided into two parts; first, the means by which the paper is rendered sensitive, and the image impressed upon it;—and secondly, the subsequent fixing and *toning*, as it may be termed, of the proof.

The present Chapter will also include, in two additional Sections, a condensed account of the most important facts relating to the properties and the mode of preservation of photographic prints.

SECTION I.

The Preparation of the Sensitive Paper.

In this Section the general theory of the preparation of Positive paper, in so far as it affects the tone and intensity of the print, will be described; the reader being referred to the second division of the Work for the formulæ required.

The Preparation of the Sensitive Paper.—The conditions which are required for producing a sharp and well defined print are—that an even layer of Chloride of Silver should exist upon the very surface of the paper, and that the particles of this Chloride should be in contact with a sufficient excess of Nitrate of Silver. These points have been already referred to at an early part of the Work (p. 19).

The material used for *sizing* the paper is of importance. English papers are usually sized with Gelatine, which is a photographic agent, and acts chemically in forming the image. Foreign papers on the other hand being sized with starch only, require an addition of Gelatine, Caseine, or Albumen, to retain the Salt at the surface of the paper, and to assist in producing the picture: if otherwise, the print will be flat and "mealy," as it is termed. Albumen especially produces a beautifully smooth surface, and is advantageously employed in printing small portraits and stereoscopic subjects.

The uniform surface distribution of the Chloride of Silver is sometimes interfered with by a faulty structure of the paper, causing it to absorb liquids unevenly, and in consequence the pictures, when removed from the printing frame, appear *spotted*. Another cause producing the same effect, is the employment of too weak a solution of Nitrate of Silver, or the removal of the sheet from the Nitrate bath before the Chloride of Ammonium has been perfectly decomposed; it is thus rendered unequally sensitive at different portions of the surface, and the prints have the characteristic marbled appearance above referred to.

A sufficient excess of Nitrate of Silver being essential, it is important to bear in mind, that the quantity of this salt eventually remaining in the paper, is much influenced by the manner in which the solution is applied. If it be laid on by *floating*, then the proportion of Nitrate to that of Chloride of Sodium should be about as 3 to 1 (the atomic weights are nearly as 5 to 2) ; but if the plan of brushing or spreading with a glass rod be adopted, 7 to 1 or 8 to 1 will not be too much.

The Darkening of the Sensitive Paper by Light.—The operator should be familiar with the changes of colour which indicate the progress of the reduction of the sensitive layer. Much in this respect depends upon the kind of organic matter used, but there is always a regular sequence of tints; in the case of a paper prepared simply with Chloride of Ammonium and Nitrate of Silver, it is as follows: pale violet, violet-blue, slate-blue, *bronze* or copper-colour. When the *bronzed* stage is reached, there is no further change. On immersion in the fixing bath of Hyposulphite, the violet tones due to Subchloride of Silver are destroyed, and the print assumes a red or brown colour, which is deepest and most intense in the parts where the light has acted longest.

Hence we see, that, to produce a good print, it is essential that the Negative should possess considerable intensity in the dark parts. Pale and feeble Negatives yield proofs which are wanting in vigour, and have a flat and indistinct appearance. The combination cannot be exposed to light for a sufficient length of time to bring about the requisite degree of reduction of the Chloride of Silver; and hence the deepest shadows of the resulting Positive are not sufficiently dark, and there is *a want of contrast* which is fatal to the effect.

A good Negative should be so opaque as to preserve the lights of the printed image beneath clear, *until the darkest shades are about to pass into the bronze or coppery condition.* If the amount of intensity be less than this, the finest effect cannot be obtained.

CONDITIONS AFFECTING THE SENSITIVENESS OF THE PAPER AND THE INTENSITY OF THE IMAGE.

Some of the principal of these are as follows:—

a. *The Strength of the Salting Bath.*—The *sensibility*

of the paper is regulated up to a certain point by the amount of salt* used in the preparation. The quantity of alkaline Chloride determines the amount of Chloride of Silver; and with a proper excess of Nitrate of Silver, papers are to a certain point more sensitive in proportion as they contain more of the Chloride.

Highly sensitized papers darken rapidly, and pass very completely into the bronze stage. Those containing less Chloride darken more slowly, and do not become bronzed with the same intensity of light. A Photographic print, formed upon paper highly salted and sensitized, is usually vigorous, with great contrast of light and shade; particularly so when the printing is conducted in a strong light. Hence it will be an advantage, with a feeble Negative, and in dull weather, to *double* the ordinary quantity of Salt, whereas in the case of an intense Negative, and with direct sunlight, the deep shadows will be too much bronzed unless the quantity of Chloride and Nitrate of Silver in the paper be kept low.

In proportion as Photographic papers are highly salted and sensitized, they become more prone to change colour spontaneously in the dark.

 b. *Proportion of Nitrate of Silver.*—The compound on which a positive print is formed is a Chloride, or an organic Salt of Silver, *with an excess of Nitrate of Silver*. Nothing is gained by increasing the proportion of Chloride of Sodium, unless at the same time an addition be made to the quantity of free Nitrate in the sensitizing Bath.

 A surface of Chloride of Silver with a bare excess of Nitrate, darkens on exposure, but it does not reach the bronzed stage; the action appearing to stop at a certain point. On placing the print in Hyposulphite of Soda, it becomes very red and pale, and when tinted, looks cold and slaty, without depth or intensity.

 * The difference in the atomic weights of the various soluble Chlorides used in salting must be borne in mind. Ten grains of Chloride of Ammonium contain as much Chlorine as eleven of Chloride of Sodium, or twenty-two grains of Chloride of Barium. (See the Vocabulary, Part III.)

c. *The sensitiveness and intensity affected by substituting the Oxide of Silver for the Nitrate.*—Many operators employ a solution of Oxide of Silver in Ammonia* or Nitrate of Ammonia, in preparing Chloride of Silver paper. By doing so, a great increase of sensitiveness, and also of intensity of image, is obtained. This will be understood if we remember that the action of light in producing the print is of a *reducing* nature. Hence the substitution of Oxide for Nitrate of Silver facilitates the decomposition; just as *Ammonio-Nitrate* of Silver is more readily reduced by Gallic or Pyrogallic Acid than the simple Nitrate (see p. 31).

Ammonio-Nitrate paper has the disadvantage of soon *discolouring* when kept; but it is very serviceable in printing during the winter months. The proportion of Chloride in the salting Bath may, if desired, be considerably reduced; the intensity of action being greatly exalted by the use of the Oxide of Silver.

d. *Employment of organic matters.*—Those recommended in this work are—Albumen, Gelatine, and Iceland Moss. Albumen adds much to the sensibility of the paper, and gives very fine surface definition. A less amount of Chloride is required than in the case of plain paper simply salted, the glutinous character of Albuminous liquids causing more of the fluid to be retained upon the surface of the paper, and the animal matter assisting the reduction. By varying the proportion of salt, both feeble and intense Negatives may be printed successfully upon albuminized paper. No process gives better results, either as regards sensitiveness, or in faithfully rendering all the finer details of the Negative, than the process with Albumen.

Iceland Moss, when boiled in water, yields a mucilaginous liquid which is conveniently employed as a vehicle for Chloride of Silver; it increases the sensitiveness of the

* The chemistry of Ammonio-Nitrate of Silver is explained in the Vocabulary, Part III.

paper and gives additional power of bronzing, by assisting to reduce the free Nitrate of Silver. Many other organic matters, tending to absorb oxygen, would act in the same way.

Gelatine is used in positive printing; it is analogous to Albumen in composition, and, like it, forms a red compound with suboxide of Silver. It is serviceable in keeping the print at the surface of the paper, but does not alter the sensibility or the general appearance of the finished picture so greatly as Albumen.

e. *Impurities in Nitrate of Silver.*—Nitrate of Silver used for Photographic printing should be free from even a trace of Protonitrate of Mercury, since it is known that the precipitation of Chloride of Mercury prevents the darkening of Chloride of Silver by light.

The peculiar condition of Nitrate of Silver spoken of at page 101, in which it is thought to contain Oxides of Nitrogen, is likely to interfere with Photographic printing. This is probably the explanation of a faulty state of the Nitrate solution, in which it yields red and feeble positives, and does not darken in colour in exciting albuminized paper. The remedy will be, to fuse the Nitrate of Silver at a moderate heat before dissolving it.

THE COLOUR OF THE IMAGE INFLUENCED BY THE PREPARATION OF THE SENSITIVE PAPER.

This subject should be studied by those who desire to print with taste. By introducing a few simple modifications into the mode of preparing the sensitive paper, almost any variety of tint may be obtained.

The tendency of the "toning" process, to which the print is afterwards to be submitted, is to darken the colour, and, if gold be used, to give a shade of *blue*. Hence, if the Positive be printed of a red tone, it will change in the gold Bath to a purple; whereas if left, after exposure to light and fixing, of a dark brown or sepia tint, it passes by toning into a pure black.

The Positive should look warm and bright on its removal from the printing frame : but the tint which remains after immersion in Hyposulphite of Soda is the proper colour of the simply fixed print.

The following points may be mentioned as affecting the colour and general appearance of the picture.

a. *The proportions of Salt and Nitrate of Silver.*— Highly salted and sensitized papers give a *darker* image than those which, containing a small proportion of Chloride of Silver, are less sensitive to light. Hence in printing upon paper weakly sensitized, in order to bring out the finer details of a highly intense negative, we find the image unusually red after fixing, and of a brown or mulberry colour when toned. The above remarks apply also in some degree to the strength of the Nitrate Bath, and especially so when no organic matter excepting Gelatine is employed,—in such a case the image will be *darker* after fixing, if the proportion of free Nitrate of Silver be large.

b. *Effect of Oxide of Silver on the colour.*—Prints formed upon Ammonio-Nitrate papers highly salted are of a sepia colour after fixing, and usually of a pure black or a purple-black when toned. With the increased facility of reduction by light afforded by use of *Oxide* of Silver, there is also less redness in the print. But if the quantity of salt used in preparing the paper be reduced to a minimum (one grain to the ounce or less), for the sake of economy or to improve the half-tone, then the usual red colour returns, and the Positive is brown or purple after toning, in place of black. Thus by employing a solution of Oxide of Silver, the operator is enabled, without the addition of organic matter, to print Positives of a pleasing variety of tint, combined with a peculiar softness and delicacy, which cannot easily be obtained with the simple Nitrate of Silver.

c. *The colour affected by organic matter.*—Albumen is coagulated by Nitrate of Silver, and forms a permanent gloss upon the paper. The sensitive albuminized paper darkens in the sun to a chocolate-brown colour, which becomes very red on immersion in the Hyposulphite. The finished prints are clear and transparent; usually of a brown tone, or with a shade of purple when the gold Bath is newly made and active; pure blacks are not easily obtained.

Iceland Moss affects the colour of the proof to a certain extent, but less than Albumen; the finished prints are nearly black if the paper is highly salted.

The Gelatinous sizing used for the English papers, and obtained by boiling hides in water, and hardening the product by an admixture of Alum, has a *reddening* influence upon reduced Silver salts, analogous to that of Albumen, or of Caseine, the characteristic animal principle of milk. Positives printed upon English paper, commonly assume some shade of brown more or less removed from black; the darker tones being more readily obtained upon the foreign papers.

Citrates and Tartrates have a marked effect upon the colour of prints. Paper prepared with Citrate, in addition to Chloride of Silver, darkens to a fine purple colour which changes to brick-red in the fixing Bath. The Positives, when toned, are usually of a violet-purple or of a bistre tint, with a general aspect of warmth and transparency.

SECTION II.

The Processes for Fixing and Toning the Proof.

This part of the operation is one to which great attention should be paid, in order to secure bright and lasting colours: it involves more of delicate chemical change than perhaps any other department of the Art.

The first point requiring explanation is the process of fixing; to which (p. 41) brief reference has already been made. The methods adopted to improve the tint of the finished picture will then be described.

CONDITIONS OF A PROPER FIXING OF THE PROOF.

This subject is not always understood by operators, and consequently they have no certain guide as to how long the prints should remain in the fixing Bath.

The time occupied in fixing will of course vary with the strength of the solution employed; but there are simple rules which may be usefully followed. In the act of dissolving the unaltered Chloride of Silver in the proof, the fixing solution of Hyposulphite of Soda converts it into Hyposulphite of Silver (p. 43), which is soluble in an *excess* of Hyposulphite of Soda. But if there be an insufficient excess,—that is, if the Bath be too weak, or the print removed from it too speedily,—then the Hyposulphite of Silver is not perfectly dissolved, and begins by degrees to *decompose*, producing a brown deposit in the tissue of the paper. This deposit, which has the appearance of yellow spots and patches, is not usually seen upon the surface of the print, but becomes very evident when it is held up to the light, or if it be split in half, which can be readily done by glueing it between two flat surfaces of deal, and then forcing them asunder.

The reaction of Hyposulphite of Soda with Nitrate of Silver.—In order to understand more fully how *decomposition* of Hyposulphite of Silver may affect the process of fixing, the peculiar properties of this salt should be studied. With this view Nitrate of Silver and Hyposulphite of Soda may be mixed in equivalent proportions, viz. about twenty-one grains of the former salt to sixteen grains of the latter, first dissolving each in separate vessels in half an ounce of distilled water. These solutions are to be added to each other and well agitated; immediately a dense deposit forms, which is Hyposulphite of Silver.

At this point a curious series of changes commences. The precipitate, at first white and curdy, soon alters in colour: it becomes canary-yellow, then of a rich orange-yellow, afterwards liver-colour, and finally black. The

K

rationale of these changes is explained to a certain extent by studying the composition of the Hyposulphite of Silver.

The formula for this substance is as follows :—

$$AgO\ S_2O_2.$$

But AgO S_2O_2 plainly equals AgS, or Sulphuret of Silver, and SO_3, or Sulphuric Acid. The acid reaction assumed by the supernatant liquid is due therefore to Sulphuric Acid, and the black substance formed is Sulphuret of Silver. The yellow and orange-yellow compounds are earlier stages of the decomposition, but their exact nature is uncertain.

The instability of Hyposulphite of Silver is principally seen when it is in an isolated state : the presence of an excess of Hyposulphite of Soda renders it more permanent, by forming a double salt, as already described.

In fixing Photographic prints, this brown deposit of Sulphuret of Silver is very liable to form in the Bath and upon the picture ; particularly so when the *temperature* is high. To obviate it, observe the following directions :—
It is especially in the reaction between *Nitrate of Silver* and Hyposulphite of Soda that the blackening is seen ; the Chloride and other *insoluble* Salts of Silver being dissolved, even to saturation, without any decomposition of the Hyposulphite formed. Hence if the print be washed in water to remove the soluble Nitrate, a very much weaker fixing Bath than usual may be employed. But if the proofs are taken at once from the printing frame and immersed in a dilute Bath of Hyposulphite (one part of the salt to six or eight of water), *a shade of brown* may often be observed to pass over the surface of the print, and a large deposit of Sulphuret of Silver soon forms as the result of the decomposition. On the other hand, with a strong Hyposulphite Bath there is little or no discoloration, and the black deposit is absent.

The print must also be left for a sufficient time in the fixing bath, or some appearance of brown patches,* visible

* The writer has noticed that when sensitive paper is *kept for some time*

by transmitted light, may occur. Each atom of Nitrate of Silver requires *three* atoms of Hyposulphite of Soda to form the *sweet and soluble double salt*, and hence, if the action be not continued sufficiently long, another compound will be formed almost tasteless and insoluble (p. 44). Even immersion in a new Bath of Hyposulphite of Soda does not fix the print when once the yellow stage of decomposition has been established. This yellow salt is insoluble in Hyposulphite of Soda, and consequently remains in the paper.

In fixing prints by Ammonia the Author has found that the same rule may be applied as in the case of Hyposulphite of Soda, viz. that if the process be not properly performed, the white parts of the print will appear *spotted* when held up to the light, from a portion of insoluble Silver Salt remaining in the paper. Prints imperfectly fixed by Ammonia are also usually brown and discoloured upon the surface of the paper.

More exact directions as to the strength of the fixing bath and the time occupied in the process, will be given in the Second Part of the Work ; at present it may be noticed only that *Albuminized* paper, from the horny nature of its surface-coating, requires a longer treatment with the Hyposulphite than the plain paper.

THE SALTS OF GOLD AS TONING AGENTS FOR PHOTOGRAPHIC PRINTS.

The Salts of Gold have been successfully applied to the improvement of the tones obtained by simply fixing the proof in Hyposulphite of Soda. The following are the principal modes followed :—

M. Le Grey's Process.—The print, having been exposed to light until it becomes very much darker than it is in-

before being used for printing, these yellow patches of imperfect fixation are very liable to occur. The Nitrate of Silver appears gradually to enter into combination with the organic matter of the *size* of the paper, and cannot then be so easily extracted by the fixing bath.

tended to remain, is washed in water to remove the excess of Nitrate of Silver. It is then immersed in a dilute solution of Chloride of Gold, acidified by Hydrochloric Acid. The effect is to reduce the intensity considerably, and at the same time to change the dark shades to a violet or bluish tint. After a second washing with water, the proof is placed in plain Hyposulphite of Soda, which fixes it and alters the tone to a pure black or a blue-black, according to the manner of preparing the paper and the time of exposure to light.

The *rationale* of the process appears to be as follows:— the Chlorine, previously combined with Gold, passes to the reduced Silver Salt; it bleaches the lightest shades, by converting them again into white Protochloride of Silver, and gives to the others a violet tint more or less intense according to the reduction. At the same time metallic Gold is deposited, the effect of which is not visible at this stage, since the same violet tint is perceived when a solution of *Chlorine* is substituted for Chloride of Gold.

The Hyposulphite of Soda subsequently employed, decomposes the violet Subchloride of Silver, and leaves the surface of a black tint, due to the Gold and the reduced Silver Salt.

M. Le Grey's process is objectionable on account of the excessive over-printing required. This however is to a great extent obviated by a modification of the process in which an *alkaline* instead of an acid solution of the Chloride is employed; one grain of Chloride of Gold is dissolved in about six ounces of water, to which are added twenty to thirty grains of the common Carbonate of Soda. The alkali moderates the violence of the action, so that the print washed with water and immersed in the Gold Bath, is less reduced in intensity, and does not acquire the same *inky* blueness. On subsequent fixing in the Hyposulphite, the tint changes from violet to a dark chocolate-brown, which is permanent.

The Tetrathionate and Hyposulphite of Gold, employed

in toning.—After the discovery of Le Grey's mode, it was proposed, as an improvement, to add Chloride of Gold to the fixing solution, so as to obviate the necessity of using two Baths. The print, in that case, although darkened considerably, is less reduced in intensity, and the same amount of over-printing is not required. The chemical changes which ensue are different from before: they may be described as follows:—

Chloride of Gold, added to Hyposulphite of Soda, is converted into Hyposulphite of Gold, Tetrathionate of Gold, and (if the Chloride of Gold be free from excess of acid) a *red compound*, containing more of the metal than either of the others, but the exact nature of which is uncertain. Each of these three Gold Salts possesses the property of darkening the print, but not to the same extent. The activity is less as the stability of the salt is greater, and hence the red compound, which is so highly unstable that it cannot be preserved many hours without decomposing and precipitating metallic Gold, is far more active than the *Hyposulphite of Gold*, which, when associated with an excess of Hyposulphite of Soda, is comparatively permanent.

When rapidity of colouring is an object it will therefore be advisable to add Chloride of Gold to the fixing Bath of Hyposulphite rather than an equivalent quantity of Sel d'or; and by dropping a little *Ammonia* into the Chloride of Gold so as to precipitate "fulminating gold"* (a compound which dissolves in Hyposulphite of Soda with considerable formation of the unstable red salt), the activity of the Bath will be promoted.

The Author explains the action of these Salts of Gold upon the Positive print as follows:—they are unstable, and contain an excess of Sulphur loosely combined; hence, when placed in contact with the image, which has an affinity for Sulphur, the existing compound is broken up,

* Read the observations on the Explosive Properties of Fulminating Gold in the Vocabulary, Part III.

and Sulphuret of Silver, Sulphuric Acid, and metallic Gold are the results. That a minute proportion of Sulphuret of Silver is formed seems certain; but the change must be superficial, as the stability of the print is very little lessened when the process is properly performed.

Sel d'Or employed as a toning agent.—This process, which was communicated to the 'Photographic Journal' by Mr. Sutton of Jersey, has been found serviceable.

The prints are first washed in water, to which is added a little Chloride of Sodium, to decompose the free Nitrate of Silver. They are then immersed in a dilute solution of "Sel d'or," or double Hyposulphite of Gold and Soda, which quickly changes the tint from red to purple without destroying any of the details or lighter shades. Lastly, the Hyposulphite of Soda is employed to fix the print in the usual way.

This process differs theoretically from the last in some important particulars. The toning solution is applied to the print *before fixing*, which experience proves to have an important influence upon the result, it having been found that when the print is previously acted upon by Hyposulphite of Soda, the rapidity of deposition of the Gold is interfered with;—thus, a dilute solution of Sel d'or colours a print rapidly, but if to this same liquid a few crystals of Hyposulphite of Soda be added, the picture becomes red and may be kept in the Bath for comparatively a long time without acquiring the purple tones.

As Hyposulphite of Soda in excess lessens the action of the Sel d'or, so on the other hand the addition of an acid increases it. The acid does not precipitate *Sulphur*, as might be expected from a knowledge of the reaction of Hyposulphite with acid bodies (p. 137), but it favours the reduction of metallic Gold. Hence it is usual to add a little Hydrochloric Acid to the toning solution of Sel d'or, to increase the rapidity and perfection of the colouring process.

THE CONDITIONS WHICH AFFECT THE ACTION OF THE
FIXING AND TONING BATH OF GOLD AND HYPOSUL-
PHITE OF SODA.

Although the process of toning Positives by Sel d'or is
very certain in its results and gives good tints, yet, as involving a somewhat greater expenditure of time and trouble, it is not at present universally adopted. The ordinary
plan of fixing and toning in one bath has been proved to
yield permanent prints if the proper precautions are observed, but it is quite necessary, in order to ensure success, that the conditions by which its action is modified
should be understood. The more important of these are
as follows :—

a. *The* AGE *of the Bath.*—When Chloride of Gold is
added to Hyposulphite of Soda, several unstable salts are
produced, which decompose by keeping. Hence the solution is very active during the first few days after mixing ;
but at the expiration of some weeks or months, if not
used, it becomes almost inert, a reddish deposit of Gold
first forming, and eventually a mixture of black Sulphuret
of Silver and Sulphur, the former of which often adheres
to the sides of the bottle in dense shining laminæ.

When the Bath is constantly kept in use there is a loss
of Gold, which, although it is less perceived than it otherwise would be, from the fact that sulphuretting principles
are formed (see next page) capable of replacing the Gold
as toning agents—yet makes the Bath work more slowly,
and hence *over-printing* is required.

b. *Presence of free Nitrate of Silver upon the surface of
the proof.*—This produces an accelerating effect, as may
be shown by soaking the print in salt and water, to convert the Nitrate into Chloride of Silver; the action then
takes place more slowly.

The free Nitrate of Silver increases the instability of
the Gold salts ; but if present in too great an excess, it is
apt to cause a decomposition of Hyposulphite of Silver,

and consequent *yellowness* in the white parts of the proof. It is therefore particularly recommended to wash the print in water before immersing it in the fixing and toning Bath.

c. *Temperature of the solution.*—In cold weather, the thermometer standing at 32° to 40°, the Bath works more slowly than usual; whereas in the height of summer, and especially in hot climates, it occasionally becomes quite unmanageable. The best temperature for operating successfully appears to be about 60° to 65° Fahrenheit; if higher than this the solutions must be employed more dilute.

d. *Addition of Iodide of Silver.*—Some operators associate *Iodide* with Chloride in the preparation of sensitive paper for printing. Another source of the same salts is the admixture of a portion of the fixing Bath used for Negatives with the Positive toning solution. The presence of Iodides in the fixing and toning Bath is injurious: when in *large excess*, they dissolve the image, or produce yellow patches of Iodide of Silver on the lights; in smaller quantity, the deposition of the Gold is hindered, and the action proceeds more slowly. Bromides and Chlorides have not the same effect.

e. *Mode of preparing the paper.*—The rapidity of toning varies with causes independent of the Bath: thus, plain paper prints are toned more quickly than prints upon albuminized paper, and the use of English paper sized with Gelatine retards the action. Foreign papers rendered sensitive with Ammonio-Nitrate tone the most quickly.

On certain states of the fixing and toning Bath which are injurious to the proofs.—The object of using the Hyposulphite Bath is to fix the proof and to tone it by means of Gold. But it is a fact familiar to the photographic chemist, that Positives can also be toned by *a sulphuretting action*, and that the colours so obtained are not very different from those which follow the employment of Gold.[*]

[*] For a more detailed account of the toning process by Sulphur, see the

Now the Hyposulphite of Soda is a substance which can be very readily made to yield up Sulphur to any bodies which possess an affinity for that element, and as the reduced Silver compound in the print has such an affinity, there is always a tendency to absorption of Sulphur when the proofs are immersed in the Bath. Consequently in many cases a sulphur toning-process is set up, and as the picture is improved by it in appearance, losing its brick-red colour and assuming a purple shade, it was at first adopted by Photographers. Experience however has shown that colours brightened in this way are less permanent than others, and are liable to fade unless kept perfectly dry. Hence the process will be discarded by all careful operators, and the object will be to avoid sulphuration as far as possible. This can be done to a great extent, and, when the Bath is properly managed, the prints will be toned almost entirely by Gold, and will, with care, be permanent.

Some of the conditions which facilitate a sulphuretting action upon the proof are as follows :—

a. *The addition of an Acid to the Bath.*—It was at one time common to add a few drops of Acetic Acid to the fixing Bath of Hyposulphite of Soda, immediately before immersing the proofs. The Bath then assumes an opalescent appearance in the course of a few minutes, and, when this milkiness is perceptible, the print begins to *tone* rapidly and becomes nearly black.

The chemical changes produced in a Hyposulphite Bath by addition of acid, may be explained thus :—The acid first *displaces* the feeble Hyposulphurous acid from its combination with Soda.

Acetic Acid   + Hyposulphite Soda.
= Acetate Soda + Hyposulphurous Acid.

Then the Hyposulphurous Acid, *not being a stable substance when isolated,* begins spontaneously to decompose,

Third Section of this Chapter, page 145. The instability of sulphuretted prints is shown in the fourth Section.

and splits up into Sulphurous Acid—which remains dissolved in the liquid, communicating the characteristic odour
of burning Sulphur—and *Sulphur*, which separates in a
finely divided state and forms a milky deposit.*

Observe therefore that free acids of all kinds must be
excluded from the fixing Bath, or, if inadvertently added,
the liquid must be set aside for some hours until the
Hyposulphurous Acid has decomposed, and, the Sulphur
having settled to the bottom, the Bath has regained its
original neutral condition.†

b. *Decomposition of the Bath by constant use.*—It has
long been known that a solution of Hyposulphite of Soda
undergoes a peculiar change in properties when much used
in fixing. When first prepared it leaves the image of a red
tone, the characteristic colour of the reduced Silver Salt,
but soon acquires the property of darkening this red colour by a subsequent communication of Sulphur. Hence
a simple fixing Bath becomes at last an active toning
bath, without any addition of Gold.

This change of properties will be found more fully explained in the abstract of the Author's researches given in
the next Section (p. 156). At present we remark only that
it is due principally to a reaction between Nitrate of Silver
and Hyposulphite of Soda, attended with decomposition of
Hyposulphite of Silver (p. 130) ; and hence, if the prints
are washed in water before immersion in the Bath, the solution will be less quickly liable to change.

Many operators state that the toning Bath having at
first been prepared with Chloride of Gold, no further addition of this substance will be required. This no doubt
is correct, but in such case the proofs will at last be toned

---

* From the Vocabulary, Part III., it will be seen that commercial Chloride
of Gold usually contains *free Hydrochloric Acid;* hence a considerable deposit of Sulphur takes place on adding it to the Hyposulphite solution,.
and the liquid must not be used immediately.

† The chemical reader will understand the decomposition of free Hyposulphurous Acid by the following equation :—$S_2O_2 = SO_2$ and S.

by Sulphur more than by Gold, and will not possess the
same stability; the Bath will also, after long use, be found
to acquire a distinct *acid* reaction to test-paper, the acidity
being due to a peculiar principle generated by decomposing
Hyposulphite of Silver, and which is shown to have an
injurious action upon the print (p. 158). To avoid this the
solution should be kept *neutral to test-paper* by means of
a drop of Ammonia, if required; and when it begins to be
exhausted, and does not tone (quickly) a print from which
the free Nitrate of Silver has been removed by washing, a
fresh quantity of Chloride of Gold should be added.

c. *Tetrathionate in the Hyposulphite Bath.*—The Author
has shown that the Tetrathionates, which are analogous
to the Hyposulphites, have an active sulphuretting action
upon Positive prints (see the papers in the next Section).
Very fine colours can be obtained in this way; but toning
by Sulphur having been proved to be wrong in principle,
the formulæ given in the first two editions of this Work
have been omitted.*

The bodies which produce Tetrathionate when added
to a solution of Hyposulphite of Soda, and hence are inadmissible in the toning process, are as follows:—Free
Iodine, Perchloride of Iron, Chloride of Copper, Acids of
all kinds (in the latter case the acid first produces Sulphurous Acid, and the Sulphurous Acid, if present in any
quantity, by reacting upon Hyposulphite of Soda, forms
Tetrathionate and Trithionate of Soda).

Chloride of Gold also produces a mixed Tetrathionate
of Gold and Soda when added to the fixing Bath (p. 133);
but as the quantity of Chloride used is small, the prints
are far less sulphuretted than in the case of toning Baths
prepared by Tetrathionate without Gold.

* The preparation of a toning bath by Tetrathionate, without Gold, is
described in the next Section, but it is not recommended for practical use.

SECTION III.

The Author's Researches in Photographic Printing.

Having been long engaged in conducting experiments upon the composition and properties of the reduced material forming the Photographic image, and especially with a view of determining the exact conditions under which the picture may be considered permanent, the Author has thought it advisable to give the results of these researches in the form of an abstract of the original papers read at the meetings of the Photographic Society.

A previous perusal of these papers will put the reader in possession of the principal facts upon which are founded the precautions advised in the next Section for the preservation of Photographic prints. In order to keep the Work as nearly as possible within its original limits, and also for the purpose of distinguishing the present Section from the others, as one referring principally to scientific details, the type has been reduced to the size of that used in the Appendix.

ON THE CHEMICAL COMPOSITION OF THE PHOTOGRAPHIC IMAGE.

The determination of the chemical nature of the Photographic image in its various forms is a point of much importance, both as indicating the conditions required for the preservation of works of art of that class, and also as a guide to the experimenter in selecting bodies likely to have an effect as chemical agents in Photography.

It has been stated by some who have given attention to the subject, that the image is formed in all cases of pure metallic Silver, and that any observable variations in its colour and properties, are due to a difference in the molecular arrangement of the particles. But this hypothesis, although involving much that is correct, yet does not contain the whole truth, for it is evident that the chemical properties of the Photographic image often bear no resemblance to those of a metal. One Photograph may also differ essentially from another, so that we are led to infer the existence of two varieties, the first of which is less of a metallic nature than the second.

In investigating the subject, the principal point appeared to be to examine the action of light upon Chloride of Silver, and afterwards to associate the Chloride with organic matter in order to imitate the conditions under which Photographs are obtained.

The following is an epitome of the conclusions arrived at:—

Action of Light upon Chloride of Silver.—The process is accompanied by a separation of Chlorine, but its product is not a mere mixture of Chloride of Silver and Metallic Silver; if it were so, we cannot suppose that the darkening would take place beneath the surface of Nitric Acid, which it is found to do. A definite Subchloride of Silver seems to be formed, the most important property of which is its decomposition by fixing agents, such as Ammonia, and Hyposulphite of Soda, both of which destroy the violet colour, dissolving out Protochloride of Silver, and leaving a small quantity of a grey residue of metallic Silver.

Inasmuch therefore as all Photographic pictures require fixing, we may conclude that if they could be produced upon pure and isolated Chloride of Silver (which however is not the case), they would consist solely of metallic Silver.

Decomposition of organic Salts of Silver by Light.—Compounds of Oxide of Silver with organic bodies, are as a rule darkened by exposure to light, but the process does not always consist in a simple reduction to the metallic state. This assertion is proved by the employment of the following tests.

a. *Mercury.*—Little or no amalgamation takes place on triturating the darkened salt with this metal.

b. *Ammonia and fixing agents.*—These usually produce only a limited amount of action. Thus, the Albuminate of Protoxide of Silver is perfectly soluble in Ammonia; but after having been reddened by exposure to light, it is little or not at all affected.

c. *Potash.*—Animal matters coagulated by Nitrate of Silver, and reduced by the sun's rays, are dissolved by boiling Potash, the solution being clear and of a blood-red colour. Metallic Silver, it is presumed, if present, would remain insoluble.

d. *Boiling Water.*—Gelatine treated with Nitrate of Silver

and exposed to light, loses its characteristic property of dissolving in hot water. This experiment is conclusive.

The above facts justify us in supposing the existence of combinations of organic matter with a low Oxide of Silver; and analysis indicates further that the relative proportion of each constituent in these compounds may vary. For instance, when Citrate of Silver is reduced by light, and acted on with Ammonia, a black powder remains, which was found to contain as much as 95 per cent. real Silver; but Albuminate of Silver treated in the same way yields on analysis less of metallic Silver, and more volatile and carbonaceous matter.

The use of *Ammonio*-Nitrate of Silver in preparing the salt tends also to increase the relative quantity of metal left in the compound after reduction and fixing. The length of time during which the light has acted, has also a modifying effect of the same kind,—the product of reduction by a powerful light being more nearly in the state of metal, and containing less both of Oxygen and organic matter.

Action of Light upon Chloride of Silver associated with organic matter.—Photographs formed on Chloride of Silver alone, would, after fixing, consist of metallic Silver, but such a process could not be carried out in practice. The addition of organic matter is absolutely necessary in order to increase the sensitiveness, and to prevent the image from being dissolved in the Bath of Hyposulphite of Soda. The blue Subchloride of Silver is decomposed by fixing, a very scanty proportion of grey metallic Silver remaining insoluble; but the red compound of Suboxide of Silver with organic matter is almost unaffected by Hyposulphite of Soda, or Ammonia.

The increase of sensitiveness and intensity produced by the use of organic matter is accompanied also by a change in the composition of the picture; the image losing the metallic character which it possesses when formed on pure Chloride of Silver, and resembling in every respect the product of the action of light upon organic Salts of Silver.

There are certain characteristic tests which may usefully be employed in distinguishing the metallic image from what may be termed the organic or non-metallic image. One of these tests is Cyanide of Potassium. An image formed upon pure Chloride of Silver, although pale and feeble, may, after fixing, be immersed in dilute solution of Cyanide of Potassium without injury. But a photograph on Chloride of Silver supported by an organic basis, is much acted upon by Cyanide of Potassium, quickly losing its finer details.

A second test is the Hydrosulphate of Ammonia. If no organic matter be employed, the image becomes darker and more intense by treatment with a soluble Sulphuret; whilst the non-metallic image, formed on an organic surface, is quickly bleached and faded. The action of Sulphur upon the image is indeed a mode of determining the real quantity of Silver present. When existing in a very finely divided layer, Sulphuret of Silver often appears yellow; but in a thicker layer it is black. Hence the colour of the Photograph, after treatment with Sulphuretted Hydrogen, is an indication of the proportion of metal present, and the reason of the organic image becoming so perfectly faded is because it contains a minimum of Silver in relation to the intensity. We see, therefore, that the addition of organic matter to Chloride of Silver does not so much increase the actual quantity of Silver reduced by light, as it adds to its opacity by associating other elements with the Silver, and altogether modifying the composition of the image.

The employment of *oxidizing agents* shows also that in an ordinary Photographic process by the direct action of light, other elements besides Silver assist in forming the image: the pictures being found to be easily susceptible of oxidation, whereas the metallic image formed on pure Chloride of Silver resists oxidation.

Composition of DEVELOPED *images.*—By exposing sensitive layers of the Iodide, the Bromide, and the Chloride of Silver to the light for a short time only, and subsequently developing with Gallic Acid, Pyrogallic Acid, and the protosalts of Iron, a variety of images may be obtained, which differ from each other materially in every important particular, and a comparison of which assists the determination of the disputed point.

The appearance and properties of the developed Photograph are found to vary with the existence of the following conditions.

1st. *The surface used to sustain the sensitive layer.*—There is a peculiarity in the image formed on *Collodion.* Collodion contains Pyroxyline, a substance which behaves towards the salts of Silver in a manner different from that of most organic bodies, exhibiting no tendency to assist their reduction by light. Hence Chloride of Silver on Collodion darkens far more slowly than the same salt upon Albumen, and the image, after fixing, is feeble and metallic. Iodide of Silver on Collodion, exposed and developed, gives usually a more metallic image, with less intensity, than Iodide of Silver upon Albumen, or on paper sized with Gelatine. By adding to the Collodion a body which has an affinity for low oxides of Silver, such for instance as Glycyrrhizine, the opacity of the developed image is increased.

2nd. *The nature of the sensitive salt.*—When *Iodide* of Silver is used to receive the latent impression, the image after development, although lacking intensity of colour by reflected light, is more nearly in the condition of metallic Silver than if Bromide or Chloride of Silver be substituted ; and of the three salts, the Chloride gives the most intensity, with the least quantity of metallic Silver. This rule applies especially when organic matters, Gelatine, Glycyrrhizine, etc., are present.

3rd. *The developing agent employed.*—An organic developing agent like Pyrogallic Acid may be expected to produce a Collodion image more intense, but less metallic, than an inorganic developer, such as the Protosulphate of Iron.

4th. *The length of time during which the light has acted.*— Over-action of the light favours the production of an image which is dark by reflection and brown or red by transmission, corresponding in these particulars to what may be termed the non-metallic image containing an oxide of Silver.

5th. *The stage of the development.*—The red image first formed on the application of the developer to a gelatinized or albuminized surface of Iodide of Silver is less metallic, and more easily injured by destructive tests, than the black image, which is the result of prolonging the action. Developed photographs which are of a bright red colour after fixing, correspond in properties to images obtained by the direct action of light on paper prepared with Chloride of Silver, more nearly than to Collodion, or even to fully developed Talbotype Negatives.

To conclude the Paper, the following may be offered in the way of recapitulation :—An image consisting of metallic silver, as a rule, reflects white light, and shows as a positive when laid on black velvet ; but a non-metallic organic image is dark, and represents the shadows of a picture. Collodion positives deve-loped with protosalts of Iron are nearly or quite metallic. Photographs on Albumen or Gelatine less so than those on Collodion. Developed Photographs contain more Silver than others, if the development has been prolonged. The half shadows of the image in a Positive Print are especially liable to suffer under injurious conditions, since they contain the Silver in a less perfect state of reduction.*

ON THE VARIOUS AGENCIES DESTRUCTIVE TO PHOTO-GRAPHIC PRINTS.

Action of Sulphuretting Compounds upon Positive Prints.— It was first noticed by Mr. T. A. Malone, that the most intense Photograph might be destroyed by acting upon it with solution of Sulphuretted Hydrogen or a soluble Sulphuret, for a sufficient length of time.

The changes produced by a sulphuretting compound acting upon the red image of a simply fixed print are these :—the colour is first darkened, and a degree of brilliancy imparted to it ; this is the effect termed "toning." Then the warm tint by degrees alters to a colder shade, the *intensity* of the whole image is lessened, and the half-tones turn yellow. Lastly, the full shadows pass also from black to yellow, and the print fades.

Now in this peculiar reaction we notice the following points of interest. If at that particular stage at which the print has reached its maximum of blackness, it be raised partially out of the liquid and allowed to project into the air, the part so treated becomes yellow before that which remains immersed. Again, if a print toned by Sulphur be placed in a pan of water to wash, after the lapse of several hours it is apt to assume a faded appearance in the half-tones. The full shadows, in which the reduced

---

* The Author omits, in this place, all mention of *molecular conditions* affecting intensity, inasmuch as at the present time nothing positive has been determined with regard to them. It is however known that in the use of the protosalts of Iron as developing agents, the appearance of the image is much influenced by the rapidity with which the reduction is effected—the particles of Silver being larger and more metallic when the development is conducted slowly. The process of electro-plating and other chemical operations of a similar kind prove that the physical properties of metals precipitated from solutions of their salts, vary greatly with the degree of fineness and arrangement of their particles.

Silver salt is thicker and more abundant, retain their black colour for a longer time, but if the action of the sulphuretting Bath be continued, every portion of the print becomes yellow.

These facts prove that *Oxygen* has an influence in accelerating the destructive action of the Sulphur compounds upon Positive prints; and this idea is borne out by the results of further experiments, for it is found that moist Sulphuretted Hydrogen has little or no effect in darkening the colour when every trace of air is excluded. When prints are washed in water they are exposed to the influence of the dissolved air which water always contains, and hence the change from black to yellow is produced.*

There are some substances which facilitate the yellow degeneration of Positives toned by Sulphur, a knowledge of which will be useful: they are—1st, powerful oxidizers, such as Chlorine, Permanganate of Potash, and Chromic Acid; these, even when highly diluted, act with great rapidity: 2nd, bodies which dissolve Oxide of Silver, as soluble Cyanides, Hyposulphites, Ammonia; also *acids* of various kinds, and hence the frequency of yellow finger impressions upon old sulphuretted prints, which are probably caused by a trace of organic (Lactic?) Acid left by contact with the warm hand.

It was at one time supposed that the Photograph in the stage at which it appears *blackened* by Sulphur, consisted of Sulphuret of Silver, and that this black Sulphuret became yellow by absorption of Oxygen and conversion into Sulphate. MM. Davanne and Girard, who examined the subject, thought that there might be two isomeric forms of Sulphuret of Silver, a black and a yellow form; the former of which passing gradually into the latter produced the fading of the impression. But neither of these views are correct; for it is proved by careful experiment, that the Sulphuret of Silver is a highly stable compound, not prone to oxidize, and, further, that the change of colour from black to yellow has no reference to a modification of this salt. The truth appears to be that the image whilst in the black stage contains other elements besides Sulphur and Silver, but when it has become yellow by the continued action of the sulphuretting compound, it is then a true Sulphuret.

---

* Further remarks upon the action of damp air upon Positives toned by Sulphur are given at p. 153.

Comparative permanence of Photographs under the action of Sulphur.—*Developed* Positives, as a rule, stand better than those printed by direct exposure to light; but much depends upon the nature of the negative process followed; and hence no general statement can be made which will not be liable to many exceptions. The mode of conducting the development must not be overlooked. The prints, which become very red in the Hyposulphite fixing Bath from the action of the developer having been stopped at too early a period, are often sulphuretted and destroyed even more readily than a vigorous sun-print obtained by direct exposure to light.

A point of even greater importance is *the nature of the sensitive surface* which receives the latent image. It is the print *developed upon Iodide of Silver* which especially resists sulphuration. In that case, not only is the preliminary toning effect of the Sulphur more slow than usual, but the impression cannot be made to fade by any continuance of the action. It loses much of its brilliancy, and is reduced in intensity, but it is not so completely destroyed as to be useless. The reason of this, as shown in the last paper, depends upon the fact that the Talbotype proofs contain the largest amount of Silver in the image.

The employment of Gold in toning does not render an ordinary sun-print as permanent as a Positive developed upon Iodide of Silver. The deep shadows of the picture are protected by the Gold, but the lighter shades not so perfectly. Hence after the Sulphur has acted, in place of the universal yellow and faded aspect presented by the simple untoned print, the Positive fully toned by Gold has black shadows with yellow half-tones. Therefore, whilst recommending the use of Gold as a toning agent, it does not seem advisable to lay too much stress upon it as a preservative from the destructive action of Sulphur.

Exposure of Positive Prints to a Sulphuretting Atmosphere. —In testing the action of a solution of Sulphuretted Hydrogen upon paper Positives, it did not appear that the conditions under which the prints were placed bore a sufficiently close resemblance to the case of Positives exposed to an atmosphere contaminated with *minute traces* of the gas; and this more particularly because it is known that *dry* Sulphuretted Hydrogen has comparatively little effect upon Photographic Prints.

The experiments were therefore repeated in a somewhat different form. A number of Positives (about three dozen) printed
in various ways, were suspended in a glass case, measuring $2\frac{1}{2}$
feet by 21 inches, and containing $7\frac{1}{2}$ cubic feet of air ; into which
was introduced, occasionally, a few bubbles of Sulphuretted Hydrogen, just sufficient to keep the air of the chamber smelling
perceptibly of the gas. A polished Daguerreotype plate was hung
up in the centre, to serve as a guide to the progress of the sulphuretting action.

By the second day the metal plate had acquired a faint yellow
hue, not easily seen except in certain positions ; but the Positives were unaffected. At the expiration of three days the majority of the pictures exhibited no signs of change, but a few untoned prints of a pale red colour, some of which had been printed
by development, and others by direct exposure to light, had perceptibly darkened.

After the eighth day, the action, appearing to progress more
slowly than at first, was stopped, and the prints removed. The
general results obtained were as follows :—

The Daguerreotype plate had become strongly tarnished with
a film of Sulphuret of Silver, which appeared yellowish-brown
in some parts and steel-blue in others. The Positives were, as
a rule, toned to a slightly colder shade, but many of them had
scarcely changed.

No obvious difference was observed between prints *developed*
on paper prepared with Chloride of Silver, and others printed by
direct exposure to light ; but in all cases the prints obtained by
those methods which give a very red image after fixing, were the
first to show the change of colour due to sulphuration, the proofs
submitted to the test having all been previously toned with Gold.

Effect of Oxidizing Agents upon Positive Prints.—It appeared
of importance to ascertain to what extent Photographic Prints
are susceptible of oxidation ; on account of the atmospheric influences to which they are necessarily exposed. In experimenting
upon this subject the following results have been obtained.

Powerful oxidizers destroy Positive Prints rapidly ; the action
usually commencing at the corners and edges of the paper, or
at any isolated point, such as a metallic speck or particle of extraneous matter, which can serve as a centre of chemical action.

This same fact is often noticed in the fading of Positives by long keeping, and therefore since other destructive actions (with the exception of that of Chlorine) do not appear to follow the same rule, it is an argument in addition to others which can be adduced, that Photographic Prints are frequently destroyed by oxidation.

Air which has been *Ozonized* by Phosphorus, and in which blue litmus-paper becomes reddened, quickly bleaches the Positive image. Oxygen gas, obtained by voltaic decomposition of acidified water and which should contain Ozone, did not appear to have an equal amount of effect, the action being comparatively slight, or altogether wanting.

Peroxide of Hydrogen obtained in solution, and in conjunction with Acetate of Baryta, by adding Peroxide of Barium to dilute Acetic Acid,* bleaches darkened Positive paper; but the effect is slow, and does not take place to a very perceptible extent if the liquid be kept alkaline to test-paper.

Nitric Acid applied in a concentrated form acts immediately upon the darkened surface, bleaching every part of the print with the exception of the bronzed shadows, which usually retain a slight residual colour. A solution of Chromic Acid is still more active. This liquid may usefully be applied to distinguish prints toned by Sulphur from others toned by Gold; the presence of metallic Gold protecting the shadows of the picture in some measure from the action of the acid. The solution should be prepared as follows :—

Bichromate of Potash . . . . . . 6 grains.
Strong Sulphuric Acid . . . . . 4 minims.
Water . . . . 12 ounces.

A solution of Permanganate of Potash is an energetic destroyer of paper positives; and, as it is a neutral substance, may conveniently be employed in testing the relative capability of withstanding oxidation possessed by different Photographic Prints. The solution should be dilute, of a pale pink hue, and the Positives must be moved occasionally, as the first effect is to decolorize a great portion of the liquid, the Permanganate oxidizing the size

* Hydrochloric Acid, which is usually recommended in place of Acetic Acid, cannot be employed in this experiment; it seems to cause a liberation of free Chlorine, which bleaches the print instantly.

and organic tissue of the paper. After an immersion of twenty
minutes to half an hour, varying with the degree of dilution, the
half-tones of the picture begin to die out, and the full shadows
become darker in colour; the bronzed portions of the print withstand the action longer, but at length the whole is changed to a
yellow image much resembling in appearance the Photograph
faded by Sulphur.

*Comparative permanence of Photographs treated with Perman-
ganate of Potash.*—Developed prints prepared by a Negative
process withstand the action better than others. But to this rule
there are exceptions; much depending upon the time of exposure to light, and the extent to which the development is carried.
Those prints which, being exposed for a short time, and afterwards strongly developed, become dark in colour and vigorous in
outline, are more permanent than others which having been over-
exposed and under-developed, lose their dark colour and become
red and comparatively faint in the Hyposulphite fixing Bath.

Positives developed upon a surface of *Chloride* of Silver on
plain paper do not resist the oxidizing action so perfectly as
those on Iodide of Silver. Prints developed upon paper prepared with Serum of Milk containing Caseine stand better than
those on plain paper.

Of prints obtained by the ordinary process of direct exposure
to light, those on plain paper are the first to fade, the oxidizing
action being most seen upon the *half-tones.* The use of *Albumen*
gives a great advantage. Developed prints on Albumen stand
far better than the same upon plain paper; and even the Albuminized sun prints are less injured by the Permanganate than
the best of the Negative prints prepared without Albumen. Caseine has the same effect, but to a less extent; and as Serum of
Milk almost invariably contains uncoagulated Caseine, its efficacy
is thus explained.

The manner of *toning* the print is a point of importance; previous sulphuration in an old Hyposulphite Bath always facilitating the oxidizing action.

Action of Chlorine upon Positive Prints.—Aqueous solution
of Chlorine destroys the Photographic image, changing it first to
a violet tint (probably Subchloride), and subsequently obliterating
it by conversion into white Chloride of Silver. The impression,

although invisible, remains in the paper, and may be developed in the form of yellow or brown Sulphuret of Silver by the action of Sulphuretted Hydrogen. It also becomes visible on exposure to light, and assumes considerable intensity if the paper be previously brushed with free Nitrate of Silver. Sulphate of Iron produces no effect upon the invisible image of Chloride of Silver; but Gallic or Pyrogallic Acid, rendered alkaline by Potash, converts it into a black deposit.

The Action of Chlorine water usually commences at the edges and corners of the print, in the same manner as that of oxidizing agents. The proofs upon Albumen are the least readily injured, and next, those developed on Iodide of Silver.

Hydrochloric Acid.—The liquid acid of sp. gr. ·116, even when free from Chlorine, acts immediately upon the half-tones of a positive print, and destroys the full shadows in the course of a few hours; a slight residual colour however usually remains in the darkest parts. The prints developed on Iodide of Silver are the most permanent.

Sulphuric, Acetic Acids, etc.—Acids of all kinds appear to exert an injurious influence upon Positive prints, and especially so upon the half-tones of the image, the effect varying with the strength of the acid and the degree of dilution with water. Even a vegetable acid like Acetic gradually darkens the colour and destroys partially or entirely the faint outlines of the picture.

Bichloride of Mercury.—The most important particulars relating to the action of this test upon Photographs are well known. The image is ultimately converted into a white powder, and hence, in the case of a Positive print, it becomes invisible; immersion in Ammonia or Hyposulphite of Soda however restores it in a form often resembling in tint the original impression. A point worthy of note is the protective effect of a deposit of Gold, which is very marked, the proof, after toning, resisting the action of the Bichloride for comparatively a long time.

Ammonia.—The effect of Ammonia upon a print is rather to *redden* the image than to destroy it; the half-tones become pale and faint, but they do not disappear. Toning with Gold enables the proof to resist the action of the strongest solution of Ammonia, and hence Ammonia may safely be employed as a fixing-agent after the use of the Sel d'or Bath.

Hyposulphite of Soda.—A concentrated solution of Hyposul-phite of Soda exercises a gradual solvent action upon the image of Photographic Prints, at the same time tending to communicate Sulphur and to darken the colour of the impression. A faint yellow outline of Sulphuret of Silver usually remains after the solution of the image is completed.

Developed prints of all kinds, but in particular the Talbotype proofs upon Iodide of Silver, are less readily dissolved by Hyposulphite of Soda than those obtained by the direct action of light. There is also a slight difference between plain and Albuminized prints, which is in favour of the former, the albuminized paper always losing somewhat more by immersion in the Hyposulphite Bath than plain Chloride paper sensitized by Nitrate of Silver.

Cyanide of Potassium.—The solvent action of Cyanide of Potassium is most energetic upon Photographs formed on paper. These images, whether developed or not, do not withstand the test so well as the impressions on Collodion. Albuminized proofs are also somewhat more easily affected than prints on simple Chloride paper sensitized with Nitrate or Ammonio-Nitrate of Silver.

Heat, moist and dry.—Long-continued boiling in distilled water has a reddening action upon Positive Prints. The image becomes at length pale and faint, resembling a print treated with Ammonia before toning. A deposit of Gold upon the image lessens, but does not altogether neutralize, the effect of the hot water. If the boiling be long continued, the violet-purple tone often imparted by the Gold invariably gives place to a chocolate-brown, which appears to be the most permanent colour. Prints *developed* by Gallic Acid upon paper prepared with Serum of Milk or with a Citrate, suffer as much as others obtained by direct action of light. Ammonio-Nitrate prints on highly salted paper, which become nearly black when toned with Gold, retain their original appearance the most perfectly; a slight diminution of brightness being the only observable difference after long boiling in water. Albumen proofs, and prints on English papers, or foreign papers prepared with Serum of Milk, Citrates, Tartrates, or any of those bodies which *redden* the reduced Salt, are, as a rule, rendered lighter in colour, and pass from purple to brown when boiled in water.

Dry heat has an opposite effect to that of hot water, usually *darkening* the colour of the image. On exposing a plain paper print simply fixed, and thoroughly freed from Hyposulphite of Soda by washing, to a current of heated air, it changes gradually from red to dark brown, in which state it continues until the temperature rises to the point at which the paper begins to char, when it resumes its original red tone, becoming at the same time faint and indistinct.

The Products of Combustion of Coal-gas a cause of Fading.— Coal-gas contains Sulphur compounds, which in combustion are oxidized into Sulphurous and Sulphuric Acids; other substances of a deleterious nature may also be present. A plate of polished silver suspended in a glass tube, through which was directed the current of heated air rising from a small gas jet, became tarnished with a white film in the course of twenty-four hours. Positive prints exposed to the same, absorbed moisture and faded; the action resembling that of oxidation, in being preceded by a general darkening in colour. Of four prints exposed, an Iodide-developed print was the least injured, and next, a print upon Albuminized paper.

ON THE ACTION OF DAMP AIR UPON POSITIVE PRINTS.

In order to ascertain this point, more than six dozen Positives, printed on every variety of paper, were mounted in new and perfectly clean stoppered glass bottles, at the bottom of each of which was placed a little distilled water, to keep the contained air always moist. They were removed at the expiration of three months, having been kept during that time, some in the dark, and others exposed to the light. As the prints were prepared by various methods, toned in different ways, and mounted with or without substances likely to exercise a deleterious action, this series of experiments will possess considerable value in determining some of the intrinsic causes of fading of Positives.*

The general results obtained were as follows :—Positives which had been *simply fixed* in Hyposulphite of Soda remained quite uninjured. Whether developed by Gallic Acid on either of the

---

* For a more detailed account of the experiments, see the original paper in the 'Photographic Journal,' vol. iii.

three Salts of Silver usually employed, or printed by direct action of light, the result was the same. Hence we may infer that the darkened material which forms the image of Photographic Prints does not readily oxidize in a damp atmosphere.

Toned Positives were found in many cases to be less permanent than Positives simply fixed. This was especially the case when the toning had been effected by *Sulphur;* all the sulphuretted prints, fixed in solution of Hyposulphite which had been long used, became yellow in the half-tones when exposed to moisture. Positives fixed and toned in Hyposulphite containing Gold were variously affected; some prepared when the solution was in an active state being unchanged, others losing a little half-tone, and others, again, fading badly. These latter were prepared in a Bath which had lost Gold and acquired sulphuretting properties; and it was noticed that they were more injured by the action of boiling water than those Positives which proved to be permanent under the influence of the moisture.

Toning by means of Chloride of Gold appeared to be highly satisfactory, but the number of prints operated upon was small. The Sel d'or process also did not injure the integrity of the image, no commencing yellowness or bleaching of half-tones being visible after exposure to the moist air.

This series of experiments confirmed the statement made in a former paper, that some tints obtained in Positive printing are more permanent than others. Violet tones produced by Sulphur invariably passed into a dull brown by the action of the moist air; and even when Gold was employed in toning, these same purple colours were usually *reddened.* This was especially the case when English papers were used, or foreign papers re-sized with Serum of Milk containing *Caseine.* The chocolate-brown tints which best stand the action of boiling water, and in particular those upon Ammonio-Nitrate paper, were least affected by the damp air; and indeed it was evident that the two agents, viz. moist air and hot water, acted alike in tending to *redden* the print, although the latter did so in the most marked manner.

It seemed also, from the results of these experiments, to be a point of great importance that the *size* should be removed from the print in order to render it indestructible by damp air. This was evidently seen in two cases where Positives, toned in an old

Hyposulphite and Gold Bath, were divided into halves, one of which was treated with a strong solution of Ammonia. The result was that the halves in which the size was allowed to remain, faded, whilst the others were comparatively uninjured. The Albumen proofs especially suffered when the size was left, in the paper, a destructive mouldiness forming, and fading the picture. The use of boiling water obviated this, and the prints so treated remained clean and bright. A partial decomposition of Albumen however occurred in some cases even when hot water was used, the gloss disappearing from the paper in isolated patches. With *Caseine* substituted for Albumen there was also a loss of half-tone; thus seeming to indicate that both these animal principles, although stable under ordinary conditions, will, even when co-agulated by Nitrate of Silver, decompose if kept long in a moist state.

The use of improper substances for mounting proved to be another determining cause of fading by oxidation. Those bodies which combine with Oxide of Silver, are likely upon theoretical grounds to destroy the half-tones of the image; and it was found, that if the picture were left in contact with Alum, Acetic Acid, etc., or with the substances which generate *an acid* by fermentation, such as paste or starch, it invariably faded.

The supposed accelerating influence of *Light* upon the fading of Positives was not confirmed by these experiments, as far as they extended. Many of the bottles containing the Photographs were placed outside the window of a house with a southern aspect during the whole of the three months with the exception of two or three weeks, but no difference whatever could be detected between Positives so treated and others kept in total darkness. It will be proper however that this part of the investigation should be repeated, allowing a longer time.

An examination of the various modes employed for coating Positives, in order to exclude the atmosphere, showed that many of them were not fitted to fulfil the purpose intended. *Waxed* prints faded quite as much when exposed to moisture as others not waxed. White wax is a substance often adulterated, and Oil of Turpentine has been shown to contain a body resembling *Ozone* in properties, and possessing the power of bleaching a dilute solution of Sulphate of Indigo. Spirit varnish applied to the surface of the picture after re-sizing with Gelatine was plainly superior to white wax, but nevertheless it did not obviate the fading effect of the moisture upon an unstable Positive which had been toned by sulphuration. Its protective influence is therefore limited.

ON THE CHANGE IN COMPOSITION WHICH HYPOSULPHITE OF SODA EXPERIENCES BY USE IN FIXING PAPER PROOFS.*

It was remarked by Photographers at an early period that the properties of the Fixing Bath of Hyposulphite of Soda became altered by constant use ; that it gradually acquired the power of *darkening* the colour of the Positive image. This change was at first referred to the accumulation of *Salts of Silver* in the Bath, and hence directions were given to dissolve a portion of blackened Chloride of Silver in the Hyposulphite in preparing a new solution.

Careful experiments performed by the Author convinced him that an error had been entertained ; since it was found that the simple solution of Chloride of Silver in Hyposulphite of Soda had no power of yielding the black tones. But it afterwards appeared that if the fixing Bath, containing dissolved Silver Salts, were set aside for a few weeks, a *decomposition* occurred in it, evidenced by the formation of a black deposit of Sulphuret of Silver; and *then* it became active in toning the proofs.

The presence of this deposit of Sulphuret of Silver indicated that a portion of Hyposulphite of Silver had spontaneously decomposed, and, knowing the products which are generated by the spontaneous decomposition of this salt, a clue to the difficulty was afforded. One atom of Hyposulphite of Silver includes the elements of one of Sulphuret of Silver and one of Sulphuric Acid. Sulphuric Acid in contact with Hyposulphite of Soda produces *Sulphurous Acid* by a process of displacement; and Plessy has shown that Sulphurous Acid reacts upon an excess of Hyposulphite of Soda, forming two of that interesting series of Sulphur compounds designated by Berzelius the "Polythionic Acids."

* These observations are condensed and re-arranged from the papers published by the Author in the 'Photographic Journal' for September and October, 1854.

It appeared therefore probable, upon theoretical grounds, that the Penta-, Tetra-, and Trithionates might produce some effect in the Hyposulphite fixing Bath. Upon making the trial these expectations were verified; and it was found that Tetrathionate of Soda added to Hyposulphite of Soda yielded a fixing and toning Bath quite equal in activity to that produced by means of Chloride of Gold.

It may be useful to review for an instant the composition of the Polythionic series of acids; it is thus represented:—

	Sulphur.	Oxygen.	Formulæ.
Dithionic or Hyposulphuric Acid	2 atoms	5 atoms	S_2O_5
Trithionic Acid	3 ,,	5 ,,	S_3O_5
Tetrathionic Acid	4 ,,	5 ,,	S_4O_5
Pentathionic Acid.	5 ,,	5 ,,	S_5O_5

The amount of *Oxygen* in all is the same, that of the other element increases progressively; hence it is at once evident that the highest member of the series might *by losing Sulphur* descend gradually until it reached the condition of the lowest.

This transition is not only theoretically possible, but there is an actual tendency to it, all the acids being unstable with the exception of the Hyposulphuric. The Alkaline Salts of these acids are more unstable than the acids themselves; a solution of Tetrathionate of Soda becomes milky in the course of a few days from deposition of Sulphur, and, if tested, is then found to contain *Tri*thionate and eventually *Di*thionate of Soda.

The cause of the change in properties of the fixing Bath being thus clearly traced to a decomposition of Hyposulphite of Silver, and a consequent generation of unstable principles capable of imparting Sulphur to the immersed proofs, it seemed desirable to continue the experiments.—

There is a peculiar *acid condition* commonly assumed by old fixing Baths, which could not be satisfactorily explained, since it was known that acids do not exist long in a free state in solution of Hyposulphite of Soda, but tend to neutralize themselves by displacing *Hyposulphurous Acid* spontaneously decomposable into Sulphurous Acid and Sulphur. This point is set at rest by the discovery of a peculiar reaction which takes place between certain salts of the Polythionic acids and Hyposulphite of Soda. A solution of Tetrathionate of Soda may be preserved for many hours unchanged; but if a few crystals of Hyposulphite of Soda be dropped in, it begins very shortly to deposit Sulphur, and continues to do so for several days. At the same time the liquid acquires *an acid reaction* to test-paper, and produces effervescence on the addition of Carbonate of Lime.

It is evident that a Sulphur acid exists which has not hitherto been described, and that this acid is formed as one of the products of the decomposition of the Hyposulphite of Silver contained in the fixing Bath. The subject is an important one to Photographers, because it is found that Hyposulphite Baths which have acquired the acid reaction, although toning quickly, yield Positives which fade on keeping. The acid may perhaps combine with the reduced Silver Salt, which, if the image be allowed to contain Suboxide of Silver, is theoretically probable.

The experiments were next directed towards ascertaining more carefully the effect of the acid fixing Bath upon the Positive proofs. Tetrathionate of Soda added to solution of Hyposulphite of Soda produces, at the expiration of twelve hours, a liquid which, when filtered from the deposited Sulphur, reddens blue litmus-paper slowly. Positive prints immersed in the Bath pass from red to black, dissolving in the half-tones, and becoming yellow and faded if the action be too long continued. On adding Carbonate of Soda in quantity sufficient to remove the acid reaction, the power of toning is much diminished, but dark colours can still be obtained by continuing the action. The solvent effect upon the half-tones, evidently caused in great measure by the acid, is lessened; whilst the tendency to yellowness in the white parts of the proof, almost disappears. These effects are more particularly manifested when the prints are immersed in the Bath immediately on their removal from the printing frame; and it is found almost impossible to preserve the whites of the impression clear, in the acid Bath, unless the Nitrate of Silver has been washed away.

Solution of half-tones and yellowness in the lights, both a source of annoyance to the operator, are thus traced in great measure to an acid condition of the fixing and toning Bath; and the remedy is obvious.

The Author's experiments upon the Tetrathionates and their re-

action with Hyposulphite of Soda likewise elicited the important
fact that *alkalies* decompose the unstable sulphuretted principle.
If the Bath be treated with Potash or Carbonate of Soda, an alkaline *Sulphuret* appears to be gradually formed, which precipitates
Sulphuret of Silver, and in the course of a few days the liquid returns to its original condition and ceases to act as a toning agent
upon the proof. The same effect takes place to a great extent
when the solution is set aside for several weeks or months; a process of spontaneous change going forward, which issues in a deposition of Sulphur and Sulphuret of Silver, and a partial loss of
sulphuretting properties in the liquid.

It may be interesting to the scientific investigator to describe
the mode of preparing a fixing and toning Bath, illustrating the
above remarks :—

Take of Nitrate of Silver . . . . 3 drachms.
Hyposulphite of Soda . . . . 4 ounces.
Water . . . . 8 ounces.

Dissolve the Nitrate of Silver in 2 ounces of the water, then from
the total quantity of Hyposulphite of Soda, weigh out

Hyposulphite of Soda . . . . 2 drachms ;

dissolve this likewise in 2 ounces of water, and the remainder of
the Hyposulphite in the other 4 ounces. Then, having the three
solutions in separate vessels, pour the Nitrate of Silver at once into
the 2-ounce solution of Hyposulphite, agitating the precipitated
Hyposulphite of Silver rapidly. In a short time it will begin to
decompose, passing from white to canary-yellow, and then to
orange-yellow. *When the orange-yellow begins to verge towards
brown*, add the 4-ounce concentrated solution of Hyposulphite,
which will at once complete the decomposition, a part of the
precipitate dissolving and the remainder becoming perfectly black.
After filtering out the black Sulphuret of Silver, the solution is
ready for use.

A Bath prepared by this formula is not usually very active, but
it shows clearly the process by which an ordinary fixing Bath may
be converted into a toning Bath by the immersion of positives
having free Nitrate of Silver upon the surface.

The following formula is more economical and gives a better

result, but it cannot be used for "Ammonio-Nitrate" prints; the addition of an alkali precipitating Sulphuret of Iron.

Strong solution of Perchloride of Iron . 6 fluid drachms.
Hyposulphite of Soda . . . . 4 ounces.
Water . . . . 8 ounces.
Nitrate of Silver . . . . 30 grains.

Dissolve the Hyposulphite of Soda in seven ounces of the water, the Nitrate of Silver in the remaining one ounce; then pour the Perchloride of Iron into the solution of Hyposulphite, by degrees, stirring all the time. The addition of the Iron Salt strikes a fine purple colour, but this soon disappears. When the liquid has become again colourless, which it does in a few minutes, add the Nitrate of Silver, stirring briskly. Perfect solution will take place without any formation of black Sulphuret.

A toning Bath prepared with Chloride of Iron will be ready for use twelve hours after mixing, but it will be more active at the expiration of a week. The solution is acid to test-paper, and *milky* from a deposit of Sulphur, which must be filtered out.

The Perchloride of Iron should be prepared by boiling Peroxide of Iron with Hydrochloric Acid, in preference to dissolving Iron wire in Aqua Regia.

The addition of the Nitrate of Silver is made in order to produce a portion of Hyposulphite of Silver in the bath; the presence of a Silver Salt having been found to modify the tint of the Positives, and to prevent their quickly turning yellow.

SECTION IV.

On the Fading of Photographic Prints.

For many years subsequent to the discovery of the process of Photographic Printing by Mr. Fox Talbot, it was not generally known that pictures so produced were easily susceptible of injury from various causes, and in particular from traces of the *fixing-agent* remaining in the paper. Hence, due care not being taken in the proper cleansing and preservation of the proofs, the majority of them faded.

This matter became at last one of such importance that the Council of the Photographic Society decided upon forming a Committee for the purpose of examining the subject. The Author was honoured by being placed upon this Committee, and the researches of which an abstract has been given in the previous Section, were undertaken at the request of the Society.

The present Section is intended to explain practically and in a concise manner the causes of the fading of Photographic Prints, and the precautions which should be taken to ensure their permanency. The *chemistry* of the subject having been fully explained in the last Section, it will suffice to refer the reader to its pages for more detailed information.

Historical evidence of the permanence of Photographs.— It is a point of interest to collect information as to the existence of old Photographs which have remained many years unchanged. There are numerous instances of Positives printed more than ten years ago, which have not perceptibly altered up to the present time. These prints are mostly on plain paper, *Albumen* not having come into use at so early a date. The general impression of practical operators however is, that fading has occurred less frequently since the introduction of Albuminized paper.

Positives printed by *development* on paper prepared by Talbot's method seem as a rule to have stood remarkably well, and instances of Talbotype Negatives having faded are rare.

Of the prints which have proved to be permanent, some are red or brown in colour, but many, being of a dark or purple shade, have evidently been *toned*, although not with Gold, the use of which was unknown to the earlier Photographers.

It is plain from data thus collected, that Photographs do not *necessarily* fade by time; and the fact that in one and the same portfolio are constantly seen prints which appear permanent, and others in an advanced state of

M

change, cannot but lead to the inference that the main causes of deterioration are *intrinsic*, depending upon some injurious matters left in the paper; which is confirmed by experiment.

Causes of fading.—The Author believes that the fading of Photographic Prints may almost invariably be referred to one or other of the following conditions :—

a. *Imperfect washing.*—This is perhaps the most important of all, and the most frequent. When Hyposulphite of Soda is allowed to remain in the paper, even in minute quantity, it gradually decomposes, with liberation of Sulphur, and destroys the print in the same way and quite as effectually as a solution of Sulphuretted Hydrogen or an alkaline sulphuret.

Imperfect washing may be suspected, if the Photograph, within a few months from the date of its preparation, *begins to get darker in colour:* the *half-tints*, which are the first to show the action, afterwards passing into the yellow stage, whilst the dark shadows remain black or brown for a longer time.

The proper mode of washing Photographs is sometimes misunderstood. The length of time during which the print lies in the water is a point of less importance, than that the water should be continually changed. When a number of Positives are placed together in a pan, and a tap turned upon them, the circulation of fluid does not necessarily extend to the bottom. This is proved by the addition of a little colouring matter, which shows that the stream flows actively above, but at the lower part of the vessel, and between the prints, there is a stationary layer of water which is of little use in washing out the Hyposulphite. Care should therefore be taken that the pictures are kept as far as possible separate from each other, and when running water cannot be had, that they are frequently moved and turned over, fresh water being constantly added. When this is done, and especially if the *size* be removed from the paper in the manner presently to be advised, *four or five hours'* washing will be sufficient. It is a mistake to allow the pictures to remain in the water for several days; which produces no good effect, and may tend to encourage a putrefactive fermentation, or the formation of a white deposit upon the image when the water contains Carbonate of Lime.

b. *Acid matters left in the Paper.*—Upon examining collections of old Photographs, it is not uncommon to find prints which are stated to have remained unaltered for a long time after their first production, but in the course of time to have lost their brilliancy, and become pale and indistinct. This kind of fading often commences at the corners and edges of the paper, and works inwards towards the centre. The Author's experiments have shown that it is principally caused by a slow process of *oxidation*.

The Photographic Image does not appear readily susceptible of oxidation unless it be previously darkened by the action of Sulphur, or placed in contact with acids or bodies which act as solvents of Oxide of Silver (p. 146). The materials often used in sizing papers, such as Alum and Resin, being of an acid nature, are directly injurious to the image; and the removal of the size, which may easily be effected by means of a dilute alkali or an alkaline carbonate, without injury to the tint, has the additional advantage of carrying out the last traces of Hyposulphite of Soda, and also the germs of *fungi*, which if allowed to remain would vegetate and produce a destructive mouldiness on exposure to damp (Chap. III. Part II.).

The fact that acids facilitate oxidation of the image is likewise a hint that Photographic Prints should not be handled too frequently, or touched with the finger·more than is necessary; the warm hand may leave behind a trace of acid* which would tend in time to produce a yellow mark.

* The writer has seen blue litmus-paper immediately reddened by being laid upon the arm of a person suffering from acute Rheumatism. This acid is probably Lactic Acid !

c. *Moisture as a cause of fading.*—Although Photographs properly printed are not readily injured by damp air (p. 153), yet as there are *impurities* of various kinds constantly floating in the atmosphere, a state of comparative dryness may be said to be essential to the preservation of all Photographs. In collecting evidence upon the subject, "wet" and "damp" are frequently alleged as having been causes of fading—the prints were hung against a damp wall during frosty weather, in a room without a fire: or the rain had been allowed to penetrate the frame! No pictures will long survive such treatment, and Photographs, like engravings and water-colour paintings, require common care to be exercised in their preservation.

d. *The modes of Mounting the Proof.*—This subject has been alluded to in the abstract of the Author's papers at p. 155. All cements which are of an acid nature, or which are liable to become *sour* by acetous fermentation, should be avoided. Flour paste is especially injurious, and many cases of fading have been traced to this cause. The addition of Bichloride of Mercury, which is often made to prevent the paste from becoming mouldy, would still more unfit it for Photographic use (p. 151). Starch is not much preferable. No substance appears better than Gelatine, which does not readily decompose, and shows no tendency to absorb atmospheric moisture. The *deliquescent* nature of many bodies is a point which should be borne in mind in mounting Photographs, and hence the use of a salt like *Carbonate of Potash*, which the writer has known to be added to paste to prevent the formation of acid, would be unadvisable.

e. *The effect of Imperfect Fixation.*—The earlier Photographers did not always succeed in properly fixing their prints, since old Photographs are often found thickly studded with spots and blotches in the tissue of the paper. These prints however are not invariably faded upon the surface, and hence it cannot be said that imperfect fixation will certainly end in the total destruction of the picture.

Still a notice of the subject may properly be introduced in this place, and the attention of the reader be once more drawn to the importance of washing the print in water on removing it from the printing frame; a decomposition invariably occurring when paper Positives *saturated with free Nitrate of Silver* are plunged in a dilute solution of Hyposulphite of Soda, containing an insufficient quantity of the salt to dissolve away the Hyposulphite of Silver before it begins to undergo spontaneous change.

f. *Exposure to an impure Atmosphere as a cause of Fading.*—The five causes of fading which precede, have mostly reference to an intrinsically faulty condition of the print. This, the sixth, explains the mode in which a Photograph carefully prepared may yet suffer injury from deleterious matters often present in the atmosphere. The air of large cities, and particularly that emanating from sewers and drains, contains Sulphuretted Hydrogen, and hence articles of silver-plate become tarnished unless placed beneath glass. The injury which a print sustains by exposure to air contaminated with Sulphuretted Hydrogen, is less than the tarnish produced upon the bright surface of a silver plate (see p. 148); but it is recommended as a precautionary measure, that Photographic Pictures be protected by glass or kept in a portfolio, and that they be not exposed too freely to the air.

The products of the combustion of coal-gas are probably more likely than the cause last named, to be a source of injury to Photographs suspended without any covering. The sulphur compounds in gas burn into Sulphurous and Sulphuric Acids, the latter of which, in combination with Ammonia, produces the sparkling crystals often observed upon the shop windows.

The question as to the manner in which the Photographic Image may best be protected from these extraneous causes of fading has been mooted, and many plans of coating prints with some impervious material have been devised. If the pictures are to be glazed or kept in a portfolio, this of itself will be sufficient, but in other cases it may perhaps be useful to apply a layer of spirit or gutta-percha varnish. The use of wax, resin, and such bodies is likely, by introducing impurities, to act injuriously rather than otherwise.

g. *Decomposition of Pyroxyline a source of Injury to Collodion Photographs.*—Collodion Positives and Negatives are usually esteemed permanent; but some have been exhibited which, having been put away in a damp place, gradually became pale and indistinct. The change commences at rough edges and isolated points, leaving the centre, as a rule, the last affected. On examination, numerous cracks are often visible, thus seeming to indicate that the Collodion film has undergone decomposition. The result of this would be the liberation of corrosive Oxides of Nitrogen, which destroy the image. Substitution compounds containing Peroxide of Nitrogen are known to be liable to spontaneous change. The bitter resin produced by acting upon white sugar with Nitro-Sulphuric Acid, if not kept perfectly dry, will sometimes evolve enough gas to destroy the cork of the bottle in which it is kept; the solution of the resin has then a strong acid reaction, and rapidly fades an ordinary Positive Print.

These facts are interesting, and indicate that Collodion Pictures, containing in themselves the elements of their destruction, should be protected from moisture by a coating of varnish.

Comparative Permanence of Photographic Prints.— There is every reason to think that the Photographic Image, however formed, is permanent, if certain injurious conditions are avoided;—in other words, that prints do not *necessarily fade*, in the same manner as fugitive colours, by a simple exposure to light and air. But supposing a case, which is the common one, of injurious influences which cannot altogether be removed, it may be useful to inquire what mode of printing gives the greatest amount of stability.

Positives produced by a short exposure to light and sub-

sequent development with Gallic Acid, may be expected
to be more permanent than ordinary sun-prints; not that
there is any reason to suppose the chemical composition
of a developed image to be peculiar, but that the use of the
Gallic Acid enables us to increase the intensity of the red
picture first formed, and to add to its stability by precipitating fresh Silver upon it. This point has not always
been attended to. It has been recommended to remove the
print from the developing solution whilst in the *red* and
early stage of development, and to produce the dark tones
subsequently by means of gold; but this plan, although
giving very good results as regards colour and gradation
of tone, appears to lessen the advantage which would otherwise accrue from the adoption of a Negative process, and
to leave the picture, as regards permanency, much in the
condition of an ordinary print obtained by direct action of
light.

The original Talbotype process, in which the latent image
is formed upon Iodide of Silver, produces, next to Collodion, the most stable image; but the difficulty of obtaining
bright and warm tints on Iodide of Silver, will stand in
the way of its adoption.

The *toning* of Paper Positives is the part of the process
which is likely to injure their stability; inasmuch as the
finest results cannot easily be obtained without incurring
sulphuration, and the action of Sulphur, if carried to any
extent, has been shown to be detrimental. The point to be
kept in view, is to alter the original structure of the image
as little as possible in toning; and it is best to use Gold
in preference to Sulphur as the colouring agent. On theoretical grounds, toning by an alkaline solution of Chloride
of Gold (p. 132), and fixing by Ammonia, is the best process; but the employment of Sel d'or, which gives a more
agreeable colour and has not been found practically to injure the image, will be generally preferred. In using *a
single fixing and toning Bath* the same object of working
by Gold rather than by Sulphur may be best attained by

maintaining the activity of the Bath by constant additions of Chloride of Gold.

The prints which are *least stable* are such as have been toned in *acid Hyposulphite Baths, without Gold;* and the difficulty of preserving such pictures from becoming yellow. in the half-tones is very great. Possibly a portion of the Sulphuretted Acid may unite with the Suboxide of Silver and cannot be removed by washing (see p. 158); but even if this be not the case, it is certain that no ordinary amount of care will obviate the occasional occurrence of fading, unless the Hyposulphite Bath be kept *neutral to test-paper.* And all those plans of toning in which Acetic or Hydrochloric Acid is mixed with Hyposulphite of Soda, and the Positive immersed whilst the liquid is in a milky state from precipitation of Sulphur, ought studiously to be avoided.

It will be well also to avoid pushing the action of the fixing and toning Bath to its utmost limits, since practice and theory both teach us that the Positives which have been long in the Hyposulphite, and consequently show a tendency to yellowness in the light parts, are most liable to lose their half-tones on keeping. Photographic Prints are found often to *darken* slightly in the course of years; and therefore by suspending the toning action at an earlier stage a margin is left for what some have termed "an improvement by time."

The use of *Albuminized* in preference to plain paper gives an advantage in protecting the image from oxidation; but if constantly exposed to moisture, a putrefactive decomposition of the animal matter may occur. The proper colour of the Albumen image being a *pale red*, the black tones should not be sought for on that variety of paper: their production, if Hyposulphite of Soda were used in toning, would probably imply an amount of Sulphuration which would more than counterbalance any advantage otherwise derivable from the Albumen.

Permanent Positives of a black colour may easily be obtained by sensitizing plain paper, free from animal matters, with Oxide of Silver in place of Nitrate. The simply fixed image being in that case of a *sepia tint*, requires a less amount of toning to change it to black. An impression was at one time prevalent that Ammonio-Nitrate prints were unstable; but so far from such being the case, they are proved to withstand the action of all destructive tests better than pictures prepared upon the same kind of paper sensitized with plain Nitrate of Silver.

Mode of testing the permanence of Positives.—The tests for Hyposulphite of Soda are not sufficiently delicate to indicate with certainty when the process of washing has been properly performed. The quantity of that salt left in the paper is usually so small and so much mixed up with organic matter, that the application of Protonitrate of Mercury or of Nitrate of Silver to the liquid which drains from the corner of the print, would probably mislead the operator.

A dilute solution of Permanganate of Potash, prepared by dissolving from half a grain or two grains of the salt, according to its purity, in one gallon of distilled water, affords a convenient mode of testing Positives as regards their power of resisting oxidation; and to an experienced eye it will prove the presence or absence of Hyposulphite of Soda, the smallest trace of which is sufficient to remove the pink colour of the Permanganate.

The most available and simple plan of testing permanence is to enclose the pictures in a stoppered glass bottle with a small quantity of water. If they retain their half-tones after a course of three months of this treatment, and do not become mouldy, the mode of printing followed is satisfactory.

Boiling water will also be found useful in distinguishing the unstable colours produced by Sulphur from those following the judicious employment of Gold; in all cases the image will at first be reddened by the hot water, but if toned without Sulphur it will, as a rule, recover much of its dark colour on drying.

The characteristic appearance of prints which have been much sulphuretted in the toning Bath, and are very liable to fade, should be known. A yellow colour in the lights is a bad sign; and if the half-tones are at all faint and indistinct, with an aspect of commencing yellowness, it is almost certain that the Positive will not last for any considerable length of time.

CHAPTER IX.

ON THE THEORY OF THE DAGUERREOTYPE AND TALBOTYPE PROCESSES, ETC.

SECTION I.

The Daguerreotype.

It was not the original intention of the Author to include a description of the Daguerreotype Process within the limits of the present Work. The Daguerreotype is a branch of the Photographic Art so distinct from the others, that, in manipulatory details, it bears very little analogy to them; a slight sketch of the theory of the process may not however be unacceptable.

All necessary remarks will fall under three heads:—The preparation of the Daguerreotype film;—the means by which the latent image is developed;—and the strengthening of the image by Hyposulphite of Gold.

The Preparation of the Daguerreotype Film.—The sensitive film of the Daguerreotypist is in many respects different from that of the Calotype or Collodiotype. The latter may be termed *wet processes,* in contradistinction to the former, where aqueous solutions are not employed. The Daguerreotype film is a pure and isolated Iodide of Silver, formed by the direct action of Iodine upon the metal. Hence it lacks one element of sensitiveness possessed by the others, viz. the presence of soluble Nitrate of Silver in contact with the particles of Iodide of Silver.

It is important to remember that the Iodide of Silver prepared by acting with vapour of Iodine upon metallic Silver, is different in its Photographic action from the yellow salt obtained by double decomposition between Iodide of Potassium and Nitrate of Silver. A Daguerreotype film, when exposed to a bright light, first darkens to an ash-grey colour and then becomes nearly white; the solubility in Hyposulphite of Soda being at the same time lessened. A Collodion film, on the other hand, if the excess of Nitrate of Silver be washed off, although it is still capable of receiving the radiant impression in the Camera, does not alter either in colour or in solubility, by exposure even to the sun's rays.

Details of the process for preparing a Daguerreotype Plate.—A copper plate of moderate thickness is coated upon the surface with a layer of pure Silver, either by the electrotype or in any other convenient manner. It is then polished with great care, until the surface assumes a brilliant metallic lustre. This preliminary operation of polishing is one of great practical importance, and the troublesome details attending it constitute one of the main difficulties to be overcome.

After the polishing is complete, the plate is ready to receive the sensitive coating. This part of the process is conducted in a peculiar manner. A simple piece of cardboard or a thin sheet of wood, previously soaked in solution of Iodine, evolves enough of the vapour to attack the silver plate; which being placed immediately above, and allowed to remain for a short time, acquires a pale violet hue, due to the formation of *an excessively delicate layer* of Iodide of Silver. By prolonging the action of the Iodine the violet tint disappears and a variety of prismatic colours are produced, much in the same way as when light is decomposed by thin plates of mica or the surface of mother-of-pearl. From violet the plate becomes of a straw-yellow, then rose-colour, and afterwards steel-grey. By continuing the exposure, the same sequence of tints is repeated; the steel-grey disappears, and the yellow and rose-colours recur. The deposit of Iodide of Silver gradually increases in thickness during these changes; but to the end it remains excessively thin and delicate. In this respect it contrasts strongly with the dense and creamy layer often employed in the Collodion process, and shows that a large proportion of the Iodide of Silver must in such a case be superfluous, as far as any influence produced by the light is concerned. An inspection of a sensitive Daguerreotype plate reveals the *microscopic* nature of the actinic changes involved in the Photographic Art, and teaches a useful lesson.

Increase of sensibility obtained by combining the joint action of Bromine and Iodine.—The original process of Daguerre was conducted with the vapour of Iodine only; but in the year 1840 it was discovered by Mr. John Goddard that the sensibility of the plate was greatly promoted by exposing it to the vapours of Iodine and Bromine in succession,—the proper time for each being regulated by the tints assumed.

The composition of this *Bromo-Iodide* of Silver, so called, is uncertain, and has not been proved to bear any analogy to that of the mixed salt obtained by decomposing a solution of Iodide and Bromide of Potassium with Nitrate of Silver. Observe also that the Bromo-Iodide of Silver is more sensitive than the simple Iodide *only when the vapour of Mercury is employed as a developer.* M. Claudet proves that if the image be formed by the direct action of light alone (see page 174), the usual condition is reversed, and that the use of Bromine under such circumstances retards the effect.

The Development and Properties of the Image.—The latent image of the Daguerreotype is developed in a manner different from that of the humid processes generally,—viz. by the action of Mercurial vapour. Mercury, or Quicksilver, is a metallic fluid which boils at 662° Fahrenheit. We are not however to suppose that the iodized plate is subjected to the vapour of Mercury at a temperature at all aproaching to 662°. The cup containing the Quicksilver is previously heated by means of a spirit-lamp to about 140°, a temperature easily borne by the hand, in most cases, without inconvenience. The amount of Mercurial vapour evolved at 140° is very small, but it is sufficient for the purpose, and after continuing the action for a short time the image is perfectly developed.

There are few questions which have given rise to greater discussion amongst chemists than the nature of the Daguerreotype image. Unfortunately, the quantity of material to be operated on is so small, that it becomes almost impossible to ascertain its composition by direct analysis. Some suppose it to consist of Mercury alone. Others have thought that the Mercury is in combination with metallic Silver. The presence of the former metal is certain, since M. Claudet shows that, by the application of a strong heat, it can actually be volatilized from the image in sufficient quantity to develope a second impression immediately superimposed.

It is a remarkable fact that an image more or less resembling that developed by Mercury can be obtained by *the prolonged action* of light alone upon the iodized plate. The substance so formed is a white powder, insoluble in solution of Hyposulphite of Soda; amorphous to the eye, but presenting the appearance of minute reflecting crystals when highly magnified. Its composition is uncertain.

For all practical purposes the production of the Daguerreotype image by light alone is useless, on account of the length of time required to effect it. This was alluded to in the third Chapter, where it was shown that in the case of the Bromo-Iodide of Silver an intensity of light 3000 times greater is required, if the use of the Mercurial vapour be omitted.

M. Ed. Becquerel's discovery of the continuing action of rays of yellow light.—Pure homogeneous yellow light has no action upon the Daguerreotype plate ; but if the iodized surface be first exposed to white light for a sufficient time to impress a latent image, and then *afterwards* to the yellow light, the action already commenced is *continued*, and even to the extent of forming the peculiar white deposit, insoluble in Hyposulphite of Soda, already alluded to.

Yellow light may therefore in this sense be spoken of as a *developing* agent, since it produces the same effect as the Mercurial vapour in bringing out to view the latent image.

A singular anomaly however requires notice, viz. that if the plate be prepared with the mixed vapours of Bromine and Iodine, in place of Iodine alone, then the yellow light cannot be made to develope the image. In fact, the same coloured ray which continues the action of white light upon a surface of *Iodide* of Silver, actually *destroys* it, and restores the particles to their original condition, with a surface of *Bromo*-Iodide of Silver.

These facts, although not of great practical importance, are interesting in illustration of the delicate and complex nature of the chemical changes produced by light.

The Strengthening of the Daguerreotype Image by means of Hyposulphite of Gold.—The use of the Hyposulphite of Gold to whiten the Daguerreotype image, and render it more lasting and indestructible, was introduced by M. Fizeau, subsequent to the original discovery of the process.

After removal of the unaltered Iodide of Silver by means of Hyposulphite of Soda, the plate is placed upon a levelling stand and covered with a solution of Hyposulphite of Gold, containing about one part of the salt dissolved in 500 parts of water. The flame of a spirit-lamp is then applied until the liquid begins to boil. Shortly a change is seen to take place in the appearance of the image; it becomes whiter than before, and acquires great force. This fact seems to prove conclusively that metallic Mercury enters into its composition, since a surface of Silver—such, for instance, as that of the Collodion image—is *darkened* by Hyposulphite of Gold.

The difference in the action of the gilding solution upon the image and the pure Silver surrounding it illustrates the same fact. This Silver, which appears of a dark colour, and forms the shadows of the image, is rendered still darker; a very delicate crust of metallic Gold *gradually* forming upon it, whereas with the image the whitening effect is immediate and striking.

SECTION II.

Theory of the Talbotype and Albumen Processes.

The Talbotype or Calotype.—This process, as practised by many at the present time, is almost identical with that originally described by Mr. Fox Talbot. The object is to obtain an even and finely divided layer of Iodide of Silver upon the surface of a sheet of paper; the particles of the Iodide being left in contact with an excess of Nitrate of Silver, and usually with a small proportion of Gallic Acid, to heighten, still further, the sensibility to light.

The English papers sized with Gelatine are commonly used for the Calotype process: they retain the film more perfectly at the surface, and the Gelatine in all probability assists in forming the image. With a foreign starch-paper, unless it be re-sized with some organic substance, the solutions sink in too deeply, and the picture is wanting in clearness and definition.

There are two modes of iodizing and sensitizing the sheets: first, by floating alternately upon Iodide of Potassium and Nitrate of Silver, in the same manner as in the preparation of papers for Positive Printing; and second, by what is termed "the single wash," which is thought by many to give superior results as regards sensitiveness and intensity of image. To iodize by this mode, the yellow Iodide of Silver, prepared by mixing solutions of Iodide of Potassium and Nitrate of Silver, is dissolved in a *strong* solution of Iodide of Potassium; the sheets are floated for an instant upon this liquid and dried; they are then re-moved to a dish of water, by the action of which the Iodide of Silver is precipitated upon the surface of the paper in a finely divided state.

The properties of a solution of Iodide of Silver in Iodide of Potassium, or of the double Iodide of Potassium and Silver, are described at page 43, a reference to which will show that the double salt is *decomposed* by a large quantity of water, with precipitation of the Iodide of Silver, this substance being insoluble in a *dilute* solution of Iodide of Potassium, although soluble in a strong solution.

Paper coated with Iodide of Silver by this mode, after proper washing in water to remove soluble salts (which if allowed to remain would attract damp), will keep good for a long time. The layer of Iodide appears of a pale prim-rose colour, and is *perfectly insensitive to light*. Even ex-posure to the sun's rays produces no change, thus indicat-ing that an excess of Nitrate of Silver is essential to the visible darkening of Iodide of Silver by light. The paper is also insensitive to the reception of an invisible image, differing in this respect from the *washed* Collodion plate, which receives an impression in the Camera, although ap-parently freed from Nitrate of Silver.

To render Calotype paper sensitive to light, it is brushed with a solution of Nitrate of Silver containing both Acetic and Gallic Acids, termed "Aceto-Nitrate" and "Gallo-Nitrate" solution. The Gallic Acid lessens the keeping qualities of the paper, but increases the sensitiveness. The Acetic Acid prevents the paper from blackening all over during the development, and preserves the clearness of the white parts; its employment is indispensable.

The paper is commonly excited upon the morning of the day upon which it is intended to be used; and the longer it is kept, the less active and certain it becomes. An ex-posure of five to eight minutes in the Camera is the ave-rage time with an ordinary view lens.

The picture is developed with a saturated solution of

N

Gallic Acid, to which a portion of Aceto-Nitrate of Silver
is added to heighten the intensity. Both Sulphate of Iron,
and Pyrogallic Acid have also been used, but they are unnecessarily strong, the invisible image being more easily developed upon paper than upon Collodion (see page 143).

After fixing the Negative by removing the unaltered
Iodide of Silver with Hyposulphite of Soda, it is well
washed and dried. White wax is then melted in with a
hot iron, so as to render the paper transparent, and to
facilitate the after-process of printing.

The Calotype cannot be compared with the Collodion
process for sensitiveness and delicacy of detail, but it possesses advantages for tourists and those who do not wish
to be encumbered with large glass plates. The principal
difficulty appears to be in obtaining a uniformly good
paper, many samples giving a speckled appearance in the
black parts of the Negative.

The Waxed Paper process of Le Grey.—This is a useful
modification of the Talbotype introduced by M. Le Grey.
The paper is waxed *before iodizing*, by which, without involving any additional operation, a very fine surface layer
of Iodide of Silver can be obtained. The Waxed Paper
Process is well adapted for tourists, from its extreme simplicity and the length of time which the film may be kept
in a sensitive condition.

Both English and foreign papers are employed : but the
former take the wax with difficulty. Mr. Crookes, who
has devoted his attention to this process, gives clear directions for waxing paper ; it is essential that pure white wax
should be obtained direct from the bleachers, since the
flat cakes sold in the shops are commonly adulterated.
The *temperature* must also be carefully kept below that
point at which decomposition of the wax takes place ; the
use of too hot an iron being a common source of failure
(see 'Photographic Journal,' vol. ii. p. 231).

The sheets of paper, having been properly waxed, are
soaked for *two hours* in a solution containing Iodide and

Bromide of Potassium, with enough free Iodine to tinge the liquid of a port-wine colour. The greasy nature of wax impedes the entry of liquids, and hence a long immersion is required. The iodizing formulæ of the French Photographers have been encumbered by the addition of a variety of substances which appear to introduce complications without giving proportional advantage, and Mr. Townshend has done the art a service by proving that the Iodide and Bromide of Potassium, with free Iodine, are sufficient. This latter ingredient was first used by Mr. Crookes; it seems to add to the clearness and sharpness of the Negatives; and as the papers are *coloured* by the Iodine, air-bubbles cannot escape detection. The process of exciting with Nitrate of Silver is also rendered more certain by the employment of free Iodine, the action of the Bath being continued until the purple colour gives place to the characteristic yellow tint of the Iodide of Silver.

Waxed Paper is rendered sensitive by immersion in a Bath of Nitrate of Silver containing Acetic Acid; the quantity of which latter ingredient should be increased when the papers are to be long kept. As the excess of Nitrate is subsequently removed, the solution may be used weaker than in the Calotype or Collodion process.

After exciting, the papers are washed with water to reduce the amount of free Nitrate of Silver to a minimum. This lessens the sensitiveness, but greatly increases the keeping qualities, and the paper will often remain good for ten days or longer.

It is a very important point, in operating with Waxed Paper, to keep the developing dishes clean. The development is conducted by immersion in a Bath of Gallic Acid containing Acetic Acid and Nitrate of Silver; and being retarded by the superficial coating of wax, there is always a tendency to an irregular reduction of Silver upon the white portions of the Negative. When the developer becomes brown and discoloured, this is almost sure to happen; and it is well known to chemists that the length

of time during which Gallic Acid and Nitrate of Silver may
remain mixed without decomposing, is much lessened by
using vessels which are dirty from having been before employed for a similar purpose. The black deposit of Silver
exercises a *catalytic* (καταλυσις, decomposition by contact)
action upon the freshly-mixed portion, and hastens its discoloration.

The Waxed Paper process is exceedingly simple and in-
expensive,—very suitable for tourists, as requiring but little
experience, and a minimum of apparatus. It is however
slow and tedious in all its stages, the sensitive papers frequently taking an exposure of twenty minutes in the Camera,
and the development extending over an hour or an hour
and a half. Several Negatives however may be developed
at the same time; and as the removal of the free Nitrate of
Silver gives the process a great advantage during hot weather, it will in all probability continue to be extensively
followed. The prints which have been sent to the Exhi-
bition of the Photographic Society, show that waxed paper
in the hands of a skilful operator may be made to delineate
architectural subjects with great fidelity, and also to give
the details of foliage and landscape Photography with distinctness.

The Albumen process upon Glass.—The process with
Albumen originated in a desire to obtain a more even surface layer of Iodide of Silver than the coarse structure
of the tissue of paper will allow. It is conducted with
simple Albumen, or "white of eggs," diluted with a convenient quantity of water. In this glutinous liquid Iodide
of Potassium is dissolved; and the solution, having been
thoroughly shaken, is set aside, the upper portion being
drawn off for use, in the same manner as in the preparation of Albuminized paper for printing.

The glasses are coated with the Iodized Albumen, and
are then placed horizontally in a box to dry. This part of
the process is considered the most troublesome, the moist
Albumen easily attracting particles of dust, and being apt

to blister and separate from the glass. If an even layer of the dried and Iodized material can be obtained, the chief difficulty of the process has been overcome.

The plates are rendered sensitive by immersion in a Bath of Nitrate of Silver with Acetic Acid added, and are then washed in water and dried. They may be kept for a long time in an excited state.

The exposure in the Camera must be unusually long; the free Nitrate of Silver having been removed by washing, and the Albumen exercising a direct retarding influence upon the sensitiveness of Iodide of Silver.

The development is conducted in the ordinary way by a mixture of Gallic Acid and Nitrate of Silver, with Acetic Acid added to preserve the clearness of the lights. It usually requires one hour or more, but may be accelerated by the gentle application of heat.

Albumen pictures are remarkable for elaborate distinctness in the shadows and minor details, and are admirably adapted for viewing in the Stereoscope; but they do not often possess the peculiar and characteristic *softness* of the Photograph upon Collodion. The process is well adapted for hot climates, being very little prone to the cloudiness and irregular reduction of Silver which are often complained of with moist Collodion under such circumstances.

M. Taupenot's Collodio-Albumen process.—This is a recent discovery which seems to involve a new principle in the Art, and gives promise of great utility.

One of the greatest objections to the Albumen process has been its want of sensitiveness; but M. Taupenot found that this was obviated to a great extent by pouring the Albumen upon a plate *previously coated with Iodide of Silver*. In this way two layers of that sensitive salt are formed, and the sensibility of the surface layer, which alone receives the image, is promoted by its resting upon a substratum of Iodide rather than upon the inert surface of the glass. In this view, if the theory be correct, the lower particle of Iodide of Silver promotes the molecular disturbance of the upper, itself remaining unchanged.

Other experimenters, pursuing the subject further, have asserted that a successful result may be obtained by coating the plate with plain Collodion and subsequently with Iodized Albumen. If this observation should prove correct, the process will be simplified and its utility increased.

In the sixth Chapter of Part II. the practical details of the Collodio-Albumen process will be described.

END OF PART I.

PART II.

PRACTICAL DETAILS OF THE COLLODION
PROCESS.

PRACTICAL DETAILS OF THE COLLODION PROCESS.

——◆——

CHAPTER I.

PREPARATION OF COLLODION.

THIS includes—the soluble Paper;—the Alcohol and Ether;—and the iodizing compounds.

The *formulæ* for Negative and Positive Collodion, and for the Nitrate Bath and developing fluids, are given in the second Chapter.

THE SOLUBLE PAPER.

Pyroxyline may be prepared either from cotton wool or from Swedish Filtering-paper. Most operators prefer the latter, from its giving a product of constant solubility, and yielding a fluid solution.* The Cotton Wool however is better adapted for use with the Sulphuric Acid and Nitre, since the Paper, from its closeness of texture, requires a longer immersion in the mixture.

Preparation of a Nitro-Sulphuric Acid of the proper strength.—There are two modes of preparing the Nitro-Sulphuric Acid: first, by mixing the acids; second, by the Oil of Vitriol and Nitre Process. The former is the

---

* Swedish filtering-paper may be procured at the operative chemists', at about five shillings the quire. Each half-sheet has the water-mark " J. H. Munktell."

best in cases where large quantities of the material are operated on, but the amateur is recommended to begin by trying the Nitre Process (p. 190) as the most simple.

PREPARATION OF NITRO-SULPHURIC ACID BY THE MIXED ACIDS.

The operator may proceed in either of two ways: first, by taking the strength of each sample of acid, and mixing according to fixed rule; second, by a more ready plan, which may be used when the exact strength of the acids is not known. Each of these will be described in succession.

a. Directions for mixing according to fixed rule.—This process is given from Mr. Hadow's original paper in the 'Quarterly Journal of the Chemical Society.' It is certain in its results if the strength of both acids be accurately determined.

A very perfect process for taking the strength of Nitric Acid is by means of powdered Marble or Carbonate of Lime, as described in various works on practical Chemistry. Sulphuric Acid may be estimated by precipitating with Nitrate of Baryta, and weighing the insoluble Sulphate with the proper precautions.

The specific gravity is not a criterion of strength to be perfectly relied on, but if it be adopted as a test, the following points must be attended to.

1st. That the temperature of the acid be at or near 60° Fahrenheit; the density of Sulphuric Acid especially is, from its small specific heat, greatly influenced by a change of temperature.

2nd. The sample of Nitric Acid must be free from Peroxide of Nitrogen, or only slightly coloured by it. This substance, when present, increases the specific gravity of the acid without adding to its available properties. A yellow sample of Nitric Acid will therefore be somewhat weaker than is indicated by the specific gravity.

3rd. The Oil of Vitriol should yield no solid residue on

evaporation. Sulphate of Lead and Bisulphate of Potash are often found in the commercial acid, and add much to its density: Oil of Vitriol containing Sulphate of Lead becomes milky on dilution.

The formula for a definite Nitro-Sulphuric Acid, of the proper strength for making the soluble Pyroxyline, may be stated thus :—

$$HO\ NO_5,\ 2\ (HO\ SO_3) + 3\tfrac{1}{2}\ HO$$

or

	Atoms.	Atomic weight.
Nitric Acid	1	54
Sulphuric Acid	2	80
Water	$6\tfrac{1}{2}$	58
		192

Having found the percentage of *real acid* which is present,* the following calculation will give the relative weights of the ingredients required to produce the formula :—

Let $\begin{cases} a = \text{percentage of real Nitric Acid,} \\ b = \quad\text{,,}\qquad\text{,,}\qquad \text{Sulphuric Acid,} \end{cases}$

then $\dfrac{5400}{a} =$ quantity of Nitric Acid,

$\dfrac{8000}{b} = \quad$,, $\quad$ Sulphuric Acid,

$192 - \dfrac{5400}{a} - \dfrac{8000}{b} = \quad$,, $\quad$ Water.

Observe that the numbers in the calculation correspond to the atomic weights recently given; and that the amount of water is derived from the *total atomic weight,* viz. 192, *minus* the sum of the weights of both acids.

Hence if the samples of acid employed are too weak for the purpose, the formula for the water gives a negative quantity.

* Tables are given in the Appendix for calculation by specific gravity; but direct analysis of the acids is the most certain.

The weight of mixed acids produced by the formula is 192 grains, which would measure somewhere about two fluid drachms. Ten times this quantity forms a convenient bulk of liquid, in which about 50 or 60 grains of Paper may be immersed.

In weighing corrosive liquids, such as Sulphuric and Nitric Acid, a small glass may be counterbalanced in the scalepan, and the acid poured in carefully. If too much is added, the excess can be removed by a glass rod, or by "the pipette" commonly employed for such a purpose.

The following example of a calculation similar to the above may be given:—

100 parts of the Oil of Vitriol $= 76\cdot65$ real acid.

    ,,   ,,   Nitric Acid $= 65\cdot4$ real acid.

therefore $\dfrac{8000}{76\cdot65} = 104\cdot3$ grains of Oil of Vitriol.

$\dfrac{5400}{65\cdot4} = 82\cdot5$,,   Nitric Acid.

$192 - 104\cdot3 - 82\cdot5 = 5\cdot2$,,   Water.

Multiplying these weights ten times, we have

Oil of Vitriol . . . 1043 grains.
Nitric Acid . . . . 825 ,,
Water . . . . . . 52 ,,

Total weight of the Nitro-Sulphuric Acid . . . } 1920 grains.

Having prepared the acid mixture of a definite strength by the above formula, the paper must be immersed according to directions given at page 191.

b. *Process for mixing Nitro-Sulphuric Acid, the strength of the two acids not having been previously determined.*— Take a *strong* sample of Nitric Acid (the yellow *Nitrous* acid, so called, succeeds well), and mix it with Oil of Vitriol as follows:—

Sulphuric Acid . . 10 fluid drachms,
Nitric Acid . . . 10   ,,

Now immerse a thermometer and note the temperature;*
it should be from 130° Fahr. to 150°. If it sinks below
120°, place the mixture in a capsule, and float upon boiling
water for a few minutes.

. A preliminary experiment with a small tuft of Cotton
Wool (cotton shows it better than paper) will then indicate
the actual strength of the Nitro-Sulphuric Acid. Stir the
tuft in the mixture for five minutes. Remove with a glass
rod, and wash with water for a short time, until no acid
taste can be perceived. If the Wool becomes *matted*, and
gelatinizes slightly on its first immersion in the acid, or if,
in the subsequent washing, the fibres appear to adhere and
to be disintegrated by the action of the water, *the Nitro-
Sulphuric Acid is too weak.* In that case add to the acid
mixture,

Oil of Vitriol, 3 drachms.

If the cotton was actually *dissolved* in the first trial, an
addition of half of a fluid ounce of Oil of Vitriol may be
required.

Supposing the cotton not to be gelatinized and to wash
well, then wring it out very dry, pull out the fibres, and
treat it in a test-tube with rectified Ether,† to which a few
drops of Alcohol have been added. If it be *insoluble*, dry
it by a gentle heat and apply a flame: a brisk explosion
indicates that the Nitro-Sulphuric Acid employed is *too
strong.* In that case, add to the twenty drachms of mixed
acids, one drachm of water, and test again, repeating the
process until a soluble product is obtained.

There is a third condition of Pyroxyline, different from
either of the above, which may be puzzling:—the fibres of

---

* In the preparation of soluble cotton, and indeed in all Photographic
manipulations, a thermometer is almost indispensable. Instruments of
sufficient delicacy for common purposes are sold in Hatton Garden and
elsewhere, at a low price. The bulb should be uncovered, to admit of being
dipped in acids, etc., without injury to the scale.

† Observe that the Ether be *pure;* if it contains too much water and
Alcohol, it will not dissolve the Pyroxyline, or will yield an opalescent solution.

the Cotton mat together very slightly or not at all on immersion, and the washing proceeds tolerably well; the compound formed is scarcely explosive, and dissolves imperfectly in Ether, leaving little nodules or hard lumps. The
ethereal solution yields, on evaporation, a film which is
opaque instead of transparent. In this case (presuming
the Ether to be good) the acid mixture is slightly too weak,
or the temperature is too low, being probably about 90°,
instead of 130° to 140° (?).

When the acid mixture has been brought to the proper
strength by a few preliminary trials, proceed according to
the directions given at the next page.

PREPARATION OF NITRO-SULPHURIC ACID BY OIL OF VITRIOL AND NITRE.

This process is recommended, in preference to the other,
to the amateur who is unable to obtain Nitric Acid of convenient strength. The common Oil of Vitriol sold in the
shops is often very good for Photographic purposes; but
it is best, if possible, to take the specific gravity, when any
doubt exists of its genuineness. At a temperature of 58°
to 60°, specific gravity 1·833 is the usual strength, and if it
falls below this, it should be rejected. (See Part III. for
'Impurities of Commercial Sulphuric Acid.')

The Nitre must be the purest sample which can be obtained. Commercial Nitre often contains a large quantity
of *Chloride of Potassium*, detected on dissolving the Nitre
in distilled water, and adding a drop or two of solution of
Nitrate of Silver. If a milkiness and subsequent curdy
deposit is formed, Chlorides are present. These Chlorides
are injurious; after the Oil of Vitriol is added, they destroy a portion of Nitric Acid by converting it into brown
fumes of Peroxide of Nitrogen, and so alter the strength
of the solution.

Nitrate of Potash is *an anhydrous salt,*—it contains
simply Nitric Acid and Potash, without any water of crystallization; still, in many cases, a little water is retained

mechanically between the interstices of the crystals, and therefore it is better to dry it before use. This may be done by laying it in a state of fine powder upon blotting-paper, close to a fire, or upon a heated metallic plate.

The sample must also be reduced to a fine powder before adding the Oil of Vitriol; otherwise portions of the salt escape decomposition.

These preliminaries having been properly observed, weigh out

Pure Nitre, powdered and dried, 600 grains.

This quantity is equivalent to 1¼ ounce Troy or Apothecaries' weight;—and to 1¼ ounce Avoirdupois weight *plus* 54 grains. Place this in a teacup or any other convenient vessel, and pour upon it.

Water . . . 1½ fluid drachms
mixed with Oil of Vitriol . 12 ,,

Stir well with a glass rod for two or three minutes, until all effervescence has ceased, and an even, pasty mixture, free from lumps, is obtained.

During the whole process, abundance of dense fumes of Nitric Acid will be given off, which must be allowed to escape up the flue or into the open air.

A modification of the formula.—The above formula will invariably succeed with a good sample of acid and pure Nitre. When tried however with Oil of Vitriol rather weaker than ordinary, and *commercial* Nitre, it may fail, the cotton being gelatinized and dissolved. When such is the case, the addition of water must be omitted or the quantity reduced from one drachm and a half to half a drachm.

GENERAL DIRECTIONS FOR IMMERSING, WASHING, AND DRYING THE PYROXYLINE.

The mixture of Sulphuric Acid and Nitre requires to be used immediately after its preparation, as it solidifies into a stiff mass on cooling; but the mixed acids may be kept for any length of time in a stoppered bottle.

When Cotton is used, the fibres should be well pulled out, and small tufts added one by one to the acid mixture, stirring with a glass rod in order to keep up a constant change of particles. The Paper is cut into squares or strips, which are introduced singly.

In either case the quantity must not be too great, or some portions will be imperfectly acted upon; about 20 grains to each fluid ounce of the mixture will be sufficient.

The *time of immersion required* varies from ten minutes with Cotton, to twenty minutes or even half an hour with the Paper. When an unusually large proportion of Sulphuric Acid is used, as in the case of a weak sample of Nitric Acid, the Cotton should be removed at the expiration of six or seven minutes, as there is a tendency to partial solution of the Pyroxyline in the acid mixture under those circumstances.

It is an advantage in some cases to prepare the material at a high temperature, but unless the proportions of the Acids are strictly according to Mr. Hadow's formula, solution of the Cotton may take place if the thermometer indicates more than 140°.

After the action is complete, the Nitro-Sulphuric Acid is left weaker than before, from addition of various atoms of water necessarily formed during the change. Hence, if the same portion be used more than once, an addition of Sulphuric Acid will be required.

Directions for Washing.—In removing the Pyroxyline from the Nitro-Sulphuric Acid, press out as much of the liquid as possible, and wash it rapidly in a large quantity of cold water, using a glass rod to preserve the fingers from injury. If it were simply thrown into a small quantity of water and allowed to remain, the rise in temperature and weakening of the acid mixture might do mischief.

The washing should be continued for at least a quarter

of an hour, or longer in the case of Paper, as it is essential
to get rid of every trace of acid. When the Nitre plan
has been adopted, a portion of the *Bisulphate of Potash*
formed adheres to the fibres, and if not carefully washed
out, an opalescent appearance is seen in the Collodion, resulting from the insolubility of this salt in the ethereal
mixture.

  If no acid taste can be perceived, and a piece of blue
litmus-paper remains in contact with the fibres for five
minutes without changing in colour, the product is thoroughly washed. It is however a safe plan to place the
Pyroxyline in running water and allow it to remain for several hours.

  Lastly, wring it out in a cloth, pull out the fibres, and
dry slowly, by a moderate heat. After drying, it may be
kept for any length of time in a stoppered bottle.

RECAPITULATION OF THE GENERAL CHARACTERS OF PY-
  ROXYLINE PREPARED IN NITRO-SULPHURIC ACID OF
  VARIOUS DEGREES OF CONCENTRATION.

  The acid mixture too strong.—The appearance of the
cotton is not much altered on its first immersion in the
mixture. It washes well, without any disintegration. On
drying, it is found to be strong in texture, and produces
a peculiar crackling sensation between the fingers, like
starch. It explodes on the application of flame, without
leaving any ash. It is insoluble in the mixture of Ether
and Alcohol, but dissolves if treated with Acetic Ether.

  The acid mixture of the proper strength.—No aggluti-
nation of the fibres of the cotton on immersion, and the
product washes well; soluble in the ethereal mixture, and
yields a *transparent* film on evaporation.

  The acid mixture too weak.—The fibres of the cotton
agglutinate, and the Pyroxyline is washed with difficulty.
On drying, the texture is found to be short and rotten. It
does not explode on being heated, but either burns quietly
with a flame, leaving behind a black ash—in which case

o

it consists simply of unaltered cotton,—or is only slightly combustible, and not explosive. It dissolves more or less perfectly in glacial Acetic Acid. When treated with the ethereal mixture, it is acted on *partially*, leaving behind lumps of unchanged cotton; the solution does not form an even transparent layer on evaporation, but becomes *opaque* and cloudy as it dries. This opacity however may be seen to a small extent with any sample of Pyroxyline, if the solvents contain too much water.

In using Swedish Paper in place of Cotton, the Pyroxyline formed in too weak a Nitrosulphuric Acid is usually insoluble in Ether and Alcohol, and burns slowly like unchanged paper.

By studying these characters, and at the same time bearing in mind that *a drachm and a half of water* in the quantities of acid given in the formula (p. 88) will suffice to cause the difference, the operator will overcome all difficulties.

PURIFICATION OF THE SOLVENTS REQUIRED FOR COLLODION.

The purity of the Ether employed is a matter of as much importance in the manufacture of a good Collodion as that of any other ingredient; this point must be attended to in order to secure a good result.

There are four kinds of Ether sold by manufacturing chemists; first, ordinary rectified Sulphuric Ether, containing a certain percentage of Alcohol and of water; specific gravity about ·750. Second, the washed Ether, which is the same agitated with an equal bulk of water, to remove the Alcohol: by this proceeding the specific gravity of the fluid is reduced considerably. Third, Ether both washed and re-rectified from a caustic alkali, so as to contain neither Alcohol nor water; in this case the specific gravity should not be higher than ·720. Fourth, "Methylated" Ether, manufactured at a lower price than the others.

Rectified Ether of 750° is not to be depended on, inasmuch as the specific gravity is often made up by adding water instead of alcohol. Methylated Ether should be used only when economy is an object, as it is prone to acidity and less certain in its properties.

Some of the qualities which render Ether unfit for Photographic purposes, are as follows:—a peculiar and disagreeable smell, either of some essential oil, or of Acetic Ether; an acid reaction to test-paper; a property of turning alcoholic solution of Iodide of Potassium brown with unusual rapidity; an alkaline reaction to test-paper; a high specific gravity, from superabundance of Alcohol and water.

The Ether which has been both washed and redistilled is always the most uniform in composition, and especially so if the second distillation be conducted from Quicklime, Carbonate of Potash, or Caustic Potash. These Alkaline substances retain the impurities, which are often of an acid nature, and leave the Ether in a fit state for use.

The redistillation of Ether is a simple process: in dealing with this fluid however the greatest caution must be exercised, on account of its inflammable nature. Even in pouring Ether from one bottle into another, if a light of any kind be near, the vapour is apt to take fire; and severe injuries have been occasioned from this cause.

Purification of Ether by redistillation from a caustic or carbonated alkali.—Take ordinary rectified Sulphuric Ether, and agitate it with an equal bulk of water to wash out the Alcohol; stand for a few minutes until the contents of the bottle separate into two distinct strata, the lower of which—*id est*, the watery stratum—is to be drawn off and rejected. Then introduce Caustic Potash, finely powdered, in the proportion of about one ounce to a pint of the washed Ether; shake the bottle again many times, in order that the water—a small portion of which is still present in solution in the Ether—may be thoroughly absorbed. Afterwards set aside for twenty-four hours (not longer, or the Potash may begin to decompose the Ether), when it will probably be observed that the liquid has become yellow, and that a flocculent deposit has formed in small quantity. Transfer to a retort of moderate capacity, supported in a saucepan of warm water, and properly connected with a condenser. On applying a gentle heat, the Ether distils over quietly, and condenses with very little loss; care must be taken that none of the alkaline. liquid contained in the body of the retort finds its way, by projection or otherwise, into the neck, so as to run down and contaminate the distilled fluid.

A more economical plan of purifying Ether is, without previous washing with water, to agitate with Carbonate of Potash or with Quicklime, and redistil at a moderate temperature.

In order to preserve Ether from decomposition, it must be kept in stoppered bottles, nearly full, and in a dark place. The stoppers should be tied over with bladder and luted, or a considerable amount of evaporation will take place, unless the neck of the bottle has been ground with unusual care. After the lapse of some months, probably a certain amount of decomposition, evidenced by the liberation of Iodine on adding Iodide of Potassium, will be found to have taken place. This however is small in amount, and not of a character to injure the fluid.

Rectification of Spirits of Wine from Carbonate of Potash.—The object of this operation is to remove a portion of water from the spirit, and so to increase its strength. Alcohol thus purified may be added to Collodion almost to any extent, without producing glutinosity and rottenness of film.

The salt termed Carbonate of Potash is a *deliquescent* salt,—that is, it has a great attraction for water; consequently when Spirits of Wine are agitated with Carbonate of Potash, a portion of water is removed, the salt dissolving in it and forming a dense liquid, which refuses to mix with the Alcohol, and sinks to the bottom. At the expiration of two or three days, if the bottle has been shaken frequently, the action is complete, and the lower stratum of fluid may be drawn off and rejected. *Pure* Carbonate of Potash is an expensive salt, and a commoner variety may be taken. It should be well dried on a heated metal plate, and powdered, before use.

The quantity may be about two ounces to a pint of spirit; or more, if an unusually concentrated Alcohol is required.

After the distillation is complete, a fluid is obtained containing about 90 per cent. of absolute Alcohol, the remaining 10 per cent. being water. The specific gravity at 60o Fahrenheit should be from ·815 to ·825; commercial Spirit of Wine being ·836 to ·840.

PREPARATION OF THE IODIZING COMPOUNDS IN A STATE OF PURITY.

These are the Iodides of Potassium, Ammonium, and Cadmium. The properties of each are more fully described in Part III.

a. *The Iodide of Potassium.*—Iodide of Potassium, as sold in the shops, is often contaminated with various impurities. The first and most remarkable is *Carbonate of Potash.* When a sample of Iodide of Potassium contains much Carbonate of Potash, it forms small and imperfect crystals, which are strongly alkaline to test-paper, and become moist on exposure to the air, from the deliquescent nature of the Alkaline Carbonate. *Sulphate of Potash* is also a common impurity; it may be detected by Chloride of Barium.

A third impurity of Iodide of Potassium is *Chloride* of Potassium; it is detected as follows :—Precipitate the salt by an equal weight of Nitrate of Silver, and treat the yellow mass with solution of Ammonia; if any Chloride of Silver is present, it dissolves in the Ammonia, and, after filtration, is precipitated in white curds by the addition of an excess of pure Nitric Acid. If the Nitric Acid employed is not pure, but contains traces of free Chlorine, the Iodide of Silver must be well.washed with distilled water before treating it with Ammonia, or the excess of free Nitrate of Silver dissolving in the Ammonia would, on neutralizing, produce Chloride of Silver, and so cause an error.

Iodate of Potash is a fourth impurity often found in Iodide of Potassium : to detect it, add a drop of dilute Sulphuric Acid, or a crystal of Citric Acid, to the solution of the Iodide; when, if much Iodate be present, the liquid will become yellow from liberation of free Iodine. The *rationale* of this reaction is as follows :—The Sulphuric Acid unites with the base of the salt, and liberates Hydriodic Acid (HI), *a colourless compound;* but if Iodic Acid (IO_5) be also present, it decomposes the Hydriodic Acid first formed, oxidizing the Hydrogen into Water (HO), and setting free the Iodine. The immediate production of a yellow colour on adding a weak acid to aqueous solution of Iodide of Potassium, is therefore a proof of the presence of an Iodate. As Iodate of Potash renders Collodion insensitive, this point should be attended to.

Iodide of Potassium may be rendered very pure by recrystallizing from Spirit, or by dissolving in strong Alcohol of sp. gr. ·823, in which Sulphate, Carbonate, and Iodate of Potash are insoluble. The proportion of Iodide of Potassium contained in saturated Alcoholic solutions varies with the strength of the spirit (*vide* Part III., article Iodide of Potassium).

Solution of Chloride of Barium is commonly used to detect impurities in Iodide of Potassium; it forms a white precipitate if Carbonate, Iodate, or Sulphate be present. In the two former cases the precipitate dissolves on the addition of *pure* dilute Nitric Acid, but in the latter it is insoluble. The commercial Iodide however is rarely so pure as to remain quite clear on the addition of Chloride of Barium.

b. *The Iodide of Ammonium.*—This salt may be prepared by adding Carbonate of Ammonia to Iodide of Iron, but more easily by the following process :—A strong solution of Hydrosulphate of Ammonia is first made, by passing Sulphuretted Hydrogen gas into Liquor Ammoniæ. To this liquid, Iodine is added until the whole of the Sulphuret of Ammonium has been converted into Iodide. When this point is reached, the solution at once colours brown from solution of free Iodine. On the first addition of the Iodine, an escape of Sulphuretted Hydrogen gas and a dense deposit of Sulphur take place. After the decomposition of the Hydrosulphate of Ammonia is complete, a portion of Hydriodic Acid—formed by the mutual reaction of Sulphuretted Hydrogen and Iodine—attacks any Carbonate of Ammonia which may be present, and causes an effervescence. The effervescence being over, the liquid is still acid to test-paper, from excess of Hydriodic Acid; it is to be cautiously neutralized with Ammonia, and evaporated by the heat of a water-bath to the crystallizing point.

The crystals should be thoroughly dried over a dish of Sulphuric Acid, and then sealed in tubes; by this means it will be preserved colourless.

Iodide of Ammonium is very soluble in Alcohol, but it is not advisable to keep it in solution, from the rapidity with which it decomposes and becomes brown.

The most common impurity of commercial Iodide of Ammonium is Sulphate of Ammonia; it is detected by its sparing solubility in Alcohol. Carbonate of Ammonia is also frequently present to a large extent, in which case an alkaline Collodion and eventually an alkaline Nitrate Bath will be produced.

c. *Iodide of Cadmium.*—This salt is formed by heating filings of metallic Cadmium with Iodine, or by mixing the two together with addition of water.

Iodide of Cadmium is very soluble both in Alcohol and Water; the solution yielding on evaporation large six-sided tables of a pearly lustre, which are permanent in the air. The commercial Iodide is sometimes contaminated with Iodide of Zinc; the crystals being imperfectly formed and slowly liberating Iodine when dissolved in Ether and Alcohol. Pure Iodide of Cadmium remains nearly or quite colourless in Collodion, if the fluid be kept in a cool and dark place.

CHAPTER II.

FORMULÆ FOR SOLUTIONS REQUIRED IN THE COL-
LODION PROCESS.

SECTION I.—Solutions for direct Positives.
SECTION II.—Solutions for Negative Photographs.

SECTION I.

Formulæ for Solutions for direct Positives.

The solutions are taken in the following order:—The
Collodion.—The Nitrate Bath.—Developing fluids.—Fixing liquids.—Whitening solution.

THE COLLODION.
Formula No. 1.

Purified Ether, sp. gr. ·720 . . . 5 fluid drachms.
Purified Alcohol, sp. gr. ·825 . . . 3 ,,   ,,
Pyroxyline . . . . 3 to 5 grains.
Pure Iodide of Cadmium or Ammo-
nium . . . . 4 grains.

Formula No. 2.

Rectified Ether, sp. gr. ·750 . . . 6 fluid drachms.
Spirits of Wine, sp. gr. ·836 . . . 2 ,,   ,,
Pyroxyline . . . . 2 to 4 grains.
Iodide of Potassium or Ammonium . 3 to 4 ,,

If the operator wishes to prepare a stock of the plain Collodion, and to iodize as required, the last formula will stand thus :—

    Rectified Ether, ·750 . . . . . . 3 fluid ounces.
    Alcohol of ·836 . . . . 2 fluid drachms.
    Pyroxyline . . . . 8 to 14 grains.

Dissolve the Pyroxyline, and let the fluid stand for forty-eight hours to subside, then draw off clear, with a siphon.

To each fluid ounce of this plain Collodion add about two fluid drachms of the following iodizing mixture :—

    Alcohol, sp. gr. ·836 . . . . . 1 fluid ounce.
    Iodide of Potassium . . . . 16 grains.

Of the two formulæ above given, the first is considered the best, but the second may be substituted for it when highly rectified spirits cannot be obtained. Iodide of Ammonium chemically pure is perhaps superior to any other Iodide for preparing a portrait Collodion, but Iodide of Cadmium, with addition of free Iodine, possesses better keeping properties, and gives very good results. A mixture of the two Iodides may also be used advantageously, or Iodide of *Potassium* may be combined with Iodide of Cadmium : this preparation has been much recommended, but the Collodion will be liable to produce a spotted film unless the salts are quite pure.

The exact quantity of Pyroxyline will vary with the temperature at which the preparation was made. The Collodion should flow smoothly on the glass and remain free from crapy lines on setting. When Iodide of Cadmium is used, the tendency to glutinosity will be a little greater than usual, which must be obviated by the directions given at page 83.

The film, after dipping in the Bath, should appear opalescent and not too yellow and creamy. Pale-blue films yield very good Positives, but with more liability to failure than thicker films (p. 109).

If the Positives are not perfectly clear and transparent in the shadows, dissolve 5 grains of Iodine in an ounce of Spirits of Wine (not methylated), and add a few drops until the Collodion assumes a golden-yellow colour.

In hot weather advantage will be gained by somewhat increasing the quantity of Alcohol in Collodion; the evaporation of the solvents being retarded, and the film rendered less liable to become dry before development. *Anhydrous* Alcohol of Sp. Gr. ·796, may be mixed with pure Ether of ·715, even to the extent of equal parts; but this is the extreme limit, and with the strongest spirit ordinarily obtainable, the Collodion will often become somewhat glutinous if the proportions (by measure) of 5 parts of Ether to 3 of Alcohol be exceeded.

Collodion prepared by Formula No. 1, and iodized with Iodide of Cadmium, may be kept for weeks or months without much loss of sensitiveness; but when Alkaline Iodides are employed as in the second Formula, Iodine is liberated, and the fluid becomes at last brown and insensitive.

THE NITRATE BATH.

Nitrate of Silver . . . . . . 30 grains.
Nitric Acid $\frac{1}{20}$ minim, or Acetic
   Acid (glacial) . . . . . . $\frac{1}{6}$ minim.
Alcohol . . . . 15 minims.
Distilled water . . . . . . 1 fluid ounce.

Nitrate of Silver which has been melted, in order to expel Oxides of Nitrogen, is always the most certain in its action: but the heat must not be raised too high or the salt will be contaminated with *Nitrite* of Silver.

In the Vocabulary (see Part III.) directions are given for the preparation and purification of Nitrate of Silver; also for the testing of distilled water, and the best substitutes when it cannot be obtained.

The Bath must be saturated with Iodide of Silver, and

Nitric Acid neutralized if it be present. Nitrate of Silver however which has undergone fusion is free from Nitric Acid.

Weigh out the total quantity of crystals of Nitrate required for the Bath, and dissolve in about two parts of water. Then take a quarter of a grain of Iodide of Potassium to each 100 grains of Nitrate, dissolve in half a drachm of water, and add to the strong solution; a yellow deposit of Iodide of Silver first forms, but on stirring is completely redissolved. When the liquid is clear, test for free Nitric Acid by dropping in a piece of blue litmus-paper. If at the expiration of two minutes the paper appears *reddened*, Nitric Acid is present, to neutralize which, add solution of Potash or Carbonate of Soda (not Ammonia) until a distinct turbidity, remaining after agitation, is produced (an excess does no harm). Then dilute down the concentrated solution with the remaining portion of the water, stirring all the time, and filter out the milky deposit. If the liquid does not at first run clear, it will probably do so on passing it again through the same filter.

Lastly, add the Acetic Acid (previously tested for impurities, see Part III.) and the Alcohol to the filtered liquid.

As the bulk of the Bath becomes lessened by use, fill it up with a solution containing 40 grains of Nitrate to the ounce, which will be found sufficient to maintain the strength nearly at the original point.

The common practice of occasionally dropping Ammonia or Potash into the solution, to remove Nitric Acid liberated by free Iodine in the Collodion, is not recommended (see p. 89).

When the Bath becomes old, and yields Positives which are highly intense or stained, and slightly foggy, with a deficiency of half-tone, it will be advisable to precipitate it with a Chloride and prepare a new one.

THE DEVELOPING FLUIDS.

Either of the three following formulæ may be used, according to the taste of the operator:—

FORMULA No. 1.

Sulphate of Iron, recrystallized . 12 to 20 grains.
Acetic Acid (glacial) . . . . 20 minims.
Alcohol . . . . 10 minims.
Water . . . . 1 fluid ounce.

FORMULA No. 2.

Pyrogallic Acid . . . . . . 2 grains.
Nitric Acid . . . . 1 drop.
Water . . . . 1 fluid ounce.

FORMULA No. 3.

Solution of Protonitrate of Iron . 1 fluid ounce.
Alcohol . . . . 20 minims.

In all these formulæ, if distilled water is not at hand, read the directions in the Vocabulary, Part III., Article "Water," for the best substitute.

Remarks upon these Formulæ.—Formula No. 1 is the most simple, since the solution can he used *as a Bath,* the same portion being employed many times successively. If it acts too rapidly, lessen the proportion of Sulphate of Iron. An addition of Nitric Acid, half a minim to the ounce, makes the image whiter and more metallic; but if too much is used, the development proceeds irregularly, and spangles of Silver are formed.

The Alcohol and Acetic Acid render the development uniform by causing the solution of Protosulphate to combine more readily with the film. The latter also has an effect in whitening the image and increasing its brightness.

Solution of Sulphate of Iron becomes red on keeping, from a gradual formation of *per*salt. When it is too weak, add more of the Protosulphate. The muddy deposit which settles to the bottom of the Bath is metallic Silver, reduced from the soluble Nitrate upon the plates.

Some operators add pure Nitrate of Potash to this developing solution, to form a *small portion* of Protonitrate of Iron. It is said to improve the colour slightly. The proportions are 10 grains of Nitrate of Potash to about 14 or 15 grains of Protosulphate of Iron.

Formula No. 2.—In this formula, if the colour of the image is not sufficiently white, try the effect of increasing the amount of Nitric Acid slightly. On the other hand, if the development is imperfect in parts, and patches of a green colour are seen, use *three grains* of Pyrogallic Acid to the ounce, with less Nitric Acid. A few drops of Nitrate of Silver solution added to the Pyrogallic, immediately before use, will augment the energy of development when blue and green spots occur.

Formula No. 3, or Protonitrate of Iron, does not require any addition of Acid; but it will be advisable, in some cases, to add to it a few drops of Nitrate of Silver immediately before developing. It gives a bright metallic image, resembling that obtained by adding Nitric Acid to Protosulphate of Iron.

The following process is commonly followed for preparing Protonitrate of Iron:—

Take of Nitrate of Baryta 300 grains;—powder and dissolve by the aid of heat in three ounces of water. Then throw in by degrees, with constant stirring, crystallized Sulphate of Iron, *powdered,* 320 grains. Continue to stir for about five or ten minutes. Allow to cool, and filter from the white deposit, which is the insoluble Sulphate of Baryta.

In place of Nitrate of Baryta, the Nitrate of Lead may be used (Sulphate of Lead being an insoluble salt), but the quantity required will be different. The atomic weights of Nitrate of Baryta and Nitrate of Lead are as 131 to 166; consequently 300 grains of the former are equivalent to 380 grains of the latter.

THE FIXING SOLUTION.

Cyanide of Potassium . . . . 2 to 12 grains.
Common Water . . . . . . 1 fluid ounce.

Cyanide of Potassium is usually preferred to Hyposulphite of Soda for fixing direct Positives; it is less liable to injure the purity of the white colour. The percentage of *Carbonate of Potash* in commercial Cyanide of Potassium is so variable that no exact directions can be given for the formula. It is best however to use it rather dilute—of such a strength that the plate is cleared gradually in from half a minute to a minute.

The solution of Cyanide of Potassium decomposes slowly on keeping, but it will usually retain its solvent power for several weeks. In order to escape inconvenience from the pungent odour evolved by this salt, many employ a vertical Bath to hold the solution; but in that case the plates must be carefully washed before fixing, as the Iron salts hasten the decomposition of the Cyanide.

THE WHITENING SOLUTION.

Bichloride of Mercury . . . 30 grains.
Distilled Water . . . . . 1 fluid ounce.

By a gentle application of heat the corrosive sublimate dissolves and forms a solution as nearly as possible *saturated* at common temperatures. The addition of a portion of Muriatic Acid enables the water to take up a larger quantity of Bichloride; but this concentrated solution, at the same time that it whitens more quickly than the other, is apt to act unequally upon different parts of the image.

Before applying the Bichloride, the image is to be fixed and the plate well washed. Either the Protosulphate of Iron or the Pyrogallic Acid with Acetic (p. 223) may be used for the development; but the whitening process is more rapid and uniform in the latter case.

SECTION II.

*Formulæ, etc., for Negative Solutions.**

THE COLLODION.

FORMULA No. 1.

Purified Ether, sp. gr. ·720 . .	5 fluid drachms.
Purified Alcohol, sp. gr. ·825 .	3 fluid drachms.
Soluble Pyroxyline	4 to 8 grains.
Pure Iodide of Cadmium or Ammonium	4 to 5 grains.

FORMULA No. 2.

Rectified Ether, sp. gr. ·750 . .	6 fluid drachms.
Alcohol, sp. gr. ·836	2 fluid drachms.
Soluble Pyroxyline	4 to 8 grains.
Iodide of Potassium or Ammonium	4 grains.

When the Collodion and Iodizing mixture are kept separate, the second formula will stand thus:—

Rectified Ether ·750	3 fluid ounces.
Alcohol of ·836	2 fluid drachms.
Pyroxyline	15 to 30 grains.

To each fluid ounce of this plain Collodion add 2 fluid drachms of the following Iodizing solution:—

Alcohol, sp. gr. ·836	1 fluid ounce.
Iodide of Potassium	20 grains.

When the temperature of the Nitro-Sulphuric Acid used in making the Pyroxyline is high (140° to 155°), it often happens that the Collodion is too fluid with 4 grains of soluble paper to the ounce, and forms a blue transparent film of Iodide on dipping the plate in the Bath. In that

---

* The same Collodion and Nitrate Bath may be used both for Positives and Negatives if required; but there are a few minor points of difference which are included in the following remarks.

case, increase the quantity of Pyroxyline from 4 grains to 6, or even to 8 grains to each ounce.

If the Collodion is glutinous, and produces a wavy surface, with less than 4 grains of Pyroxyline to the ounce, it is probable that the Alcohol is too weak, or that the soluble Cotton is badly made.

If *flakes of Iodide of Silver* are seen loose upon the surface of the film, and falling away into the Bath, the Collodion is over-iodized, and it will be impossible to obtain a good picture.

After the Collodion has been employed to coat a number of plates, the relative proportions of Alcohol and Ether contained in it become changed, from the superior volatility of the latter fluid : when it ceases to flow readily, and gives a more dense film than usual, thin it down by the addition of a little rectified Ether.

In dissolving the Pyroxyline, any fibrous or flocculent matter which resists the action of the Ether, must be allowed to subside, the clear portion being decanted for use. The Iodide of Potassium is to be *finely powdered*, and digested with the spirit until dissolved ; it is better not to apply any heat. Both Iodide of Ammonium and Iodide of Cadmium dissolve almost immediately, if the salts are pure.

The Collodion must be kept in a cool and dark place. When prepared with Iodide of Ammonium or Potassium it becomes at length high coloured and insensitive. The free Iodine may then be removed by a strip of pure zinc or silver foil,

When sensitiveness is not an object, many prefer working with an old, coloured Collodion, finding that it gives more intensity. It has been shown at page 97 that a peculiar change takes place in Collodion after iodizing, by which the intensity of the image is increased.

Directions for using Glycyrrhizine in Collodion.—The action of this material has been described at page 114. The Collodion should be iodized with the Iodide of Cadmium

P

only, or with a mixture of the Iodides and Bromides of the alkalies. The condition which calls for the employment of Glycyrrhizine is that often found in a newly made and rather glutinous Collodion, viz. sensitiveness of film, with good half-tones, but insufficient intensity in the high lights. Dissolve the Glycyrrhizine in Alcohol (not Methylated) in the proportion of 5 grains to the ounce : this solution may perhaps keep unchanged for three or four months. To each ounce of the Collodion add from one to four drops, and expose in the Camera a few seconds longer than before. The effect of the Glycyrrhizine upon the Collodion may not be fully produced immediately ; if so, the fluid must be set aside for twenty-four hours.

Use of Nitro-glucose in Collodion.—Nitro-glucose is a substance analogous to Pyroxyline, but more unstable. When added to Collodion iodized with the alkaline Iodides, it slowly decomposes, liberates Iodine, lessens the sensitiveness to a certain extent, and confers intensity. Like Glycyrrhizine, it may be used to remedy feebleness of the image, and to give opacity to the blacks. Prepare the Nitro-glucose by the directions given in the Vocabulary, Part III. Dissolve twenty grains in an ounce of *pure* spirit, and agitate with powdered chalk to remove free acid. Add from five to eight drops to each ounce of Collodion. In a few days, more or less, according to temperature, the Collodion will deepen in colour, and will be found on trial to produce a more vigorous picture.

Collodion for hot Climates.—In this case the Iodide of Ammonium should be avoided, as unstable and prone to change colour. Iodide of Cadmium may be substituted, which has been shown to remain quite colourless when dissolved in Alcohol and Ether.

Collodion iodized with the Iodide of *Potassium* will usually keep for about six weeks or two months; but no certain rule can be given, much depending upon the condition of the Ether and the heat of the weather.

Plain Collodion may retain its properties unimpaired for

five or six months, sometimes much longer; but there is a tendency to a formation of the acid principle (p. 85); and hence, on the addition of an alkaline Iodide to old Collodion, the coloration is commonly very rapid. The structure of the transparent film may also be injured by keeping plain Collodion for too long a time.

Photographers who wish to operate with Collodion in hot climates will find it advantageous to carry with them the prepared Pyroxyline and the spirituous solvents, observing that the bottles are carefully *luted*, and that a bubble of air is left in the neck of each, to allow for the necessary expansion, which might otherwise burst the glass or force out the stopper.

THE NITRATE BATH.

This solution may be prepared by the same formula as that given for direct Positives at page 203, acidifying the solution with Acetic Acid in preference to Nitric Acid.

THE DEVELOPING SOLUTION.

Pyrogallic Acid . . . . . . 1 grain.
Acetic Acid (glacial) . . . . . 10 to 20 minims,
    or Beaufoy's Acetic Acid fort. 1 fluid drachm.
Alcohol . . . . 10 minims.
Distilled Water . . . . . 1 fluid ounce.

In place of Distilled Water, pure Rain-Water may be used (see Part III., Art. "Water").

The quantity of Acetic Acid required will vary with the strength of the Acid and the temperature of the atmosphere. An excess enables the manipulator to cover the plate more easily before the action begins, but when the picture is taken in a dull light, is apt to give a bluish, inky hue to the image. In cold weather, use less of the Acetic and twice the quantity of Pyrogallic Acid. With Collodion prepared from Spirits nearly anhydrous, and iodized with Iodide of Cadmium, the full quantity of Acetic Acid will be required, as there is sometimes a little difficulty in making the developer flow up to the edge of the film.

If the image cannot be rendered sufficiently black, two or three minims of the Nitrate Bath solution may be added to each drachm towards the end of the development.

If the solution be kept for some time after its first preparation, it becomes brown and discoloured. In this state it will still develope the image, but is less likely to give a clear and vigorous picture. A solution of Pyrogallic Acid in Acetic Acid will keep for many weeks, and may be diluted down when required for use.

The following is a good formula:—

  Pyrogallic acid . . . 12 grains.
  Beaufoy's Acetic acid . 1 fluid ounce.

To one drachm add seven drachms of water.

THE FIXING LIQUID.

  Cyanide of Potassium . 2 to 12 to 20 grains.
  Water . . . . . . 1 fluid ounce.

  or, Hyposulphite of Soda . $\frac{1}{2}$ ounce.
  Water . . . . . . 1 fluid ounce.

For remarks on the Cyanide of Potassium Fixing Bath, see the last Section, page 207.

CHAPTER III.

MANIPULATIONS OF THE COLLODION PROCESS.

THESE may be classed under five heads:—Cleaning the Plates.—Coating with Iodide of Silver.—Exposure in the Camera.—Developing the image.—Fixing the image.—In addition to this, the present Chapter will include in separate Sections directions for the choice and management of lenses, for copying engravings, manuscripts, etc., and for taking stereoscopic and microscopic photographs.

CLEANING THE GLASS PLATES.

Care should be taken in selecting glass for use in Photography. The ordinary window-glass is inferior, having scratches upon the surface, each of which may cause an irregular action of the developing fluid; and the squares are seldom flat, so that they are apt to be broken in compression during the printing process.

The patent plate answers better than any other description of glass; but if it cannot be procured, the "flatted crown glass" may be substituted.

Before washing the glasses, each square should be roughened on the edges by means of a file or a sheet of emery-paper; or more simply, by drawing the edges of two plates across each other. If this precaution be omitted, the fingers are liable to injury, and the Collodion film may contract and separate from the sides.

In cleaning glasses, it is not sufficient, as a rule, to wash them simply with water; other liquids are required to remove *grease*, if present. A cream of Tripoli powder and Spirits of Wine, with a little Ammonia added, is commonly employed. A tuft of cotton is dipped in this mixture, and the glasses are well rubbed with it for a few minutes. They are then rinsed in plain water and wiped dry with a cloth.

The cloths used for cleaning glasses should be kept expressly for that purpose; they are best made of a material sold as fine "diaper," and very free from flocculi and loosely-adhering fibres. They are not to be washed *in soap and water*, but always in pure water or in water containing a little Carbonate of Soda.

After wiping the glass carefully, complete the process by polishing with an old silk handkerchief, avoiding contact with the skin of the hand. Some object to *silk*, as tending to render the glass electrical, and so to attract particles of dust, but in practice no inconvenience will be experienced from this source.

Before deciding that the glass is clean, hold it in an angular position and *breathe* upon it. The importance of attending to this simple rule will be at once seen by referring to the remarks made at page 39. In the Honey preservative and Collodio-albumen processes it is especially needful that the glasses should be thoroughly cleaned, on account of the tendency which the film has to become loosened or to blister during the development and washings. Caustic Potash, sold by the druggists under the name of "Liquor Potassæ," is very efficacious, or in place of it, a warm solution of "washing Soda" (Carbonate of Soda). Liquor Potassæ, being a caustic and alkaline liquid, softens the skin and dissolves it; it must therefore be diluted with about four parts of water and applied to the glass by means of a cylindrical roll of flannel. After wetting both sides thoroughly, allow the glass to stand for a time until several have been treated in the same way; then wash with water and rub dry in a cloth.

The use of an alkaline solution is usually sufficient to clean the glass, but some plates are dotted on the surface with small white specks, not removable by Potash. These specks may consist of hard particles of *Carbonate of Lime*, and when such is the case they dissolve readily in a dilute acid,—Oil of Vitriol, with about four parts of water added, or dilute Nitric Acid.

The objection to the use of Nitric Acid is, that if allowed to come in contact with the dress, it produces stains which cannot be removed unless *immediately* treated with an alkali. A drop of Ammonia should be applied to the spot before it becomes yellow and faded.

When Positives are to be taken, it is advisable to use additional care in preparing the glass, and especially so with pale transparent films and neutral Nitrate Bath.

After a glass has been once coated with Collodion, it is not necessary in cleaning it a second time to use anything but pure water; but if the film has been allowed to harden and become dry, possibly dilute Oil of Vitriol or Cyanide of Potassium may be required to remove stains.

When glasses have been repeatedly used in photography they often become at length so dull and stained, that it is better to reject them.

COATING THE PLATE WITH THE COLLODIO-IODIDE OF SILVER.

This part of the process, with that which follows, must be conducted in a room from which chemical rays of light are excluded. It is inferred therefore that the operator has provided himself with an apartment of that kind.

The most simple plan of preparing the room is to nail a treble thickness of yellow calico completely over the window, or a part of it, the remainder being darkened. To this a single thickness of a waterproof material made by coating linen with gutta-percha may be added as a further security against the entrance of white light, the smallest pencil of which admitted into the room would cause fogging.

It is often convenient to illuminate by means of a candle screened by yellow glass. A dark orange yellow, approaching to brown, is more impervious to chemical rays than a lighter canary yellow. Lamps suitable for the purpose are sold by the manufacturers of apparatus and chemicals.

Before coating the plate with Collodion, see that the fluid is perfectly clear and transparent, and that all particles have settled to the bottom; also that the neck of the bottle is free from hard and dry *crusts,* which, if allowed to remain, would partially dissolve and produce striæ upon the film. In taking small portraits and stereoscopic subjects, these points are of especial importance, and every picture will be spoiled if they are not attended to.

A useful piece of apparatus for clearing Collodion is that represented in the following woodcut.

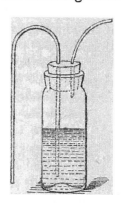

The Collodion, having been iodized some hours previously, is allowed to settle down and become clear in this bottle; then by gently blowing at the point of the shorter tube, the small glass siphon is filled, and the fluid drawn off more closely than could be done by simply pouring from one bottle to another.

When the Collodion is properly cleared from sediment, the operator takes a glass plate, previously cleaned, and wipes it gently with a silk handkerchief, in order to remove any particles of dust which may have subsequently collected. If it be a plate of moderate size, it may be held by the corners in a horizontal position, between the fore-finger and thumb of the left hand. The Collodion is to be poured on steadily until a circular pool is formed, extending nearly to the edges of the glass.

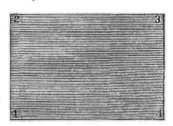

By a slight inclination of the plate the fluid is made to flow towards the corner marked 1, in the above diagram, until it nearly touches the thumb by which the glass is held : from corner 1 it is passed to corner 2, held by the forefinger; from 2 to 3, and lastly, the excess poured back into the bottle from the corner marked No. 4. It is then to be held vertically over the bottle for a moment, until it *nearly* ceases to drip, and then, by raising the thumb a little, the direction of the plate is changed, so as to cause the diagonal lines to coalesce and produce a smooth suface. The operation of coating a plate with Collodion must not be done hurriedly, and nothing is required to ensure success but steadiness of hand and a sufficiency of the fluid poured in the first instance upon the plate.

In coating larger plates, the *pneumatic* holder, which fixes itself by suction, will be found the most simple and useful.

The Proper Time for immersing the Film in the Bath.— After exposing a layer of Collodion to the air for a short time, the greater part of the Ether evaporates, and leaves the Pyroxyline in a state in which it is neither wet nor dry, but receives the impression of the finger without adhering to it. Photographers term this *setting*, and when

it takes place it is a sign that the time has come for submitting it to the action of the Bath.

If the film be lowered into the Nitrate before it has set,
the effect is the same as that produced by adding *Water*
to Collodion. The Pyroxyline is precipitated in part, and
consequently there are cracks, and the developer will not
always run up to the edge of the film. On the other hand,
if it be allowed to become too dry, the Iodide of Silver
does not form perfectly, and the film, on being washed and
brought out to the light, exhibits a peculiar iridescent appearance, and is paler in some parts than in others.

No rule can be given as to the exact time which ought
to elapse: it varies with the temperature of the atmosphere, and with the proportions of Ether and of Pyroxyline; thin Collodion containing but little Alcohol requiring
to be immersed more speedily. Twenty seconds in the
common way, or ten seconds in hot weather, will be found
an average time.

When the plate is ready, rest it upon the glass dipper,
Collodion side uppermost, and lower it into the solution by
a slow and steady movement: if any pause be made, a horizontal line corresponding to the surface of the liquid will
be formed. Then place the cover upon the vertical trough*
and darken the room, if this has not already been done.
As the presence of white light does no injury to the plate
previous to its immersion in the Bath, it is not necessary
to exclude it during the time of coating with Collodion.

When the plate has remained in the solution about twenty
seconds, lift it partially out two or three times, in order to
wash away the Ether from the surface. An immersion of
one minute to a minute and a half will usually be sufficient;
or two minutes in cold weather, and with Collodion containing but little Alcohol. Continue to move the plate
until the liquid flows off in a uniform sheet, when the de-

---

* Troughs made of gutta-percha, glass, or porcelain are commonly used;
the latter are the best, being quite opaque and not liable to cracks or
leaking.

composition may be considered to be sufficiently perfect. The principal impediment in this part of the process lies in the difficulty with which Ether and Water mix together, which causes the Collodion surface on its first immersion to appear *oily* and covered with streaks. By gentle motion the Ether is washed away, and a smooth and homogeneous layer obtained.

The plate is next removed from the dipper, and held vertically in the hand for a few seconds upon blotting paper, to drain off as much as possible of the solution of Nitrate of Silver.* It is then wiped on the back with filtering-paper, placed in a clean and dry slide, and is ready for the Camera.

The amateur is strongly recommended not to proceed to take pictures in the Camera until by a little practice he has succeeded in producing a perfect film which is uniform in every part and will bear inspection when washed and brought out to the light.

It should, if properly prepared, present the following appearance: — Smooth and uniform, both by reflected and transmitted light; free from wavy lines or markings such as would be caused by a glutinous Pyroxyline, and from opaque dots due to small particles of dust or Iodide of Silver in suspension in the Collodion.

The evidences of a *too rapid immersion in the Bath* are sought for on the side of the plate from which the Collodion was poured off. This part remains wet longer than the other, and always suffers the most; horizontal cracks or marks resembling vegetation are seen, each of which would cause an irregular action of the developing fluid. On the other hand, *the upper part* of the plate must be examined for the pale colour characteristic of a film which had *become too dry* before immersion, since the Collodion is thinner at that point than at any other.

---

* This blotting-paper must be frequently changed, or stains will be produced at the lower edge of the plate during the development.

EXPOSURE OF THE PLATE IN THE CAMERA.

After the plate has been rendered sensitive, it should be exposed and developed with all convenient despatch; the intensity of the Negatives being, with some Collodion, materially lessened by neglecting this point (see p. 100).

Ascertain that the joints of the Camera are tight in every part—that the sensitive plate, when placed in the slide, falls precisely in the same plane as that occupied by the ground glass—and that the chemical and visual foci of the Lens accurately correspond.*

Supposing the case of a portrait, next proceed to arrange the sitter as nearly as possible in a vertical position, that every part may be equidistant from the lens. Then, an imaginary line being drawn from the head to the knee, point the Camera slightly *downwards*, so that it may stand at right angles to the line. If this point be neglected the figure will be liable to be distorted in a manner presently to be shown (p. 228).

In order to succeed well with portraits, the sitter should be illuminated by an even, diffused light falling horizontally. A vertical light causes a deep shadow on the eyes and makes the hair appear grey: it must therefore be cut off by a curtain of blue or white calico suspended over the head. The direct rays of the sun are generally to be avoided, as causing too great a contrast of light and shade. This is a point on which the operator must exercise his judgment. With a feeble Collodion, a better Negative picture may often be obtained by placing the sitter quite in the open air, but when the Collodion and Bath are in the condition for giving great intensity of image, the gradation of tone will be inferior unless the light be prevented from falling too strongly upon the face and hands.

In focussing the object, cover the head and the back part of the Camera with a black cloth, and shift the Lens

* See the Second Section of this Chapter.

gently until the greatest possible amount of distinctness is obtained. Then insert the sensitive plate, and having raised the door of the slide, cover all with a black cloth during the exposure, as a security against white light finding entrance at any part excepting through the Lens.

With regard to the proper time for the exposure, so much depends upon the brightness of the light and the nature of the Collodion, that it must be left almost entirely to experience. The following general rules however may be of use :—

In a tolerably bright day in the spring or summer months, and with a newly-mixed Collodion, allow four seconds for a Positive Portrait, and eight seconds for a Negative. With a double-combination Lens of large aperture and short focus, perhaps three seconds, and six seconds, or even less, may be sufficient.

In the dull winter months, in the smoky atmosphere of large cities, or when using an old Collodion brown from free Iodine, multiply these numbers three or four times, which will be an approximation to the exposure required. It is by the appearance presented under the influence of the developer, which will immediately be described, that the operator ascertains the proper time for exposure to light.

THE DEVELOPMENT OF THE IMAGE.

The details of developing the latent image differ so much in the case of Positive and Negative pictures, that it is better to describe the two separately.

The development of direct Positives.—With Sulphate of Iron as a developer, it is most simple to develope the image by immersion. The solution may conveniently be poured into a vertical trough, such as that used for exciting, and the plate immersed by means of a glass dipper in the usual way. Unless the weather be cold, the image makes its appearance in three or four seconds, and the film is then immediately washed with clear water. Whilst in the Bath, the plate is kept in gentle motion, and the operator must not expect to see the image very distinctly, except the high lights; the shadows, being faint, are partially concealed by the unaltered Iodide, but they come out during the fixing. The action of the Sulphate of Iron is stopped at an early period, or an excess of development will be incurred. The Bath may be used repeatedly.

In using Pyrogallic Acid or Nitrate of Iron to develope glass Positives, the plate may be placed upon a levelling-stand, or held in the hand, or by the pneumatic holder, and the solution poured on quickly at one corner; by blowing gently or inclining the hand, as the case may be, it is scattered evenly over the film before the development commences.

If any difficulty is experienced in covering a plate evenly with a strong developer before the action commences, it may be overcome by using a shallow *cell* formed by cementing two or three thicknesses of window-glass on a piece of patent plate to the depth of a quarter of an inch. The size of the cell should be only slightly larger than the plate intended to be developed, that the waste of fluid may be as little as possible.

The cell is held in the left hand, and the plate being placed in it, a sufficient quantity of the developer is poured on at one corner. By a slight inclination, the fluid is caused to flow in a uniform sheet over the surface of the film, backwards and forwards. The image starts out quickly, and the developer is then at once poured off, and the film washed as before.

It is very important in developing Positives to use a sufficient quantity of the solution to cover the plate easily; otherwise oily stains and marks are formed, from the developer not combining properly with the surface of the film. For a plate five inches by four, three or four drachms will be required, and so in proportion for larger sizes.

The appearance of the Positive image after developing, as a guide to the proper time of exposure.—When the plate has been developed, it is washed, fixed, and laid upon a dark ground, such as a piece of black velvet, for inspection.

In the case of a portrait, if the features have an un-naturally black and gloomy appearance, the dark portions of the drapery, etc.; being invisible, the picture has been *under-exposed*.

On the other hand, in an over-exposed plate, the face is usually pale and white, and the drapery misty and indistinct. Much however in this respect depends upon the dress of the sitter (see p. 66), and the manner in which the light is thrown; if the upper part of the figure is shaded too much, the face may perhaps be the last to be seen. The operator should accustom himself to expend pains in the preliminary focussing upon the ground glass, and to ascertain at that time that every part of the object is equally illuminated. For this reason, pictures taken in a room are seldom successful; the light falls entirely upon one side, and hence the shadows are dark and indistinct.

The development of Negative Pictures.—This process differs in most respects from that of Positives. In the latter case, there is a tendency to over-develope the image; but in the former, to stop the action at too early a period; hence it is common to find Negative Pictures which are insufficiently developed, and too pale to print well.

In developing Negatives, many operators place the plate upon a levelling-stand, and distribute the fluid by blowing gently upon the surface; others prefer holding it in the hand and pouring the fluid on and off from a glass measure. The quantity of developer required will be less than that used for Positives, inasmuch as, if the Acetic Acid be present in sufficient excess, it is easy to cover the plate before the action begins. Some Collodion however, especially the glutinous kind, seems to repel the developer and prevent it from running up to the edge of the plate. When this is the case, or when oiliness and stains are produced, from the Bath being old and containing Ether, Alcohol must be added to the solution of Pyrogallic Acid.

With ordinary Negative Collodion, an addition of Nitrate of Silver to the developer will often be required; but the Pyrogallic Acid is to be used alone until the image has reached its maximum of intensity, which it will do in a minute or so, according to the temperature of the developing room. The plate may then be examined leisurely by placing it in front of, and at some distance from, a sheet of white paper. If it is not sufficiently black, add about four drops of the Nitrate Bath to each drachm of developer, stir well with a glass rod, and continue the action until the requisite amount of intensity is obtained. When there is any disposition in the plate to *fog* towards the end of the development, it may be obviated by fixing with Cyanide of Potassium (not Hyposulphite), and then, after a careful washing, intensifying with Pyrogallic Acid and Nitrate of Silver in the usual way. The glass which contains the mixture of Pyrogallic Acid and Nitrate of Silver must be washed out after each plate, as the black deposit hastens the discoloration of the fresh solution (p. 179).

Appearance of the Negative image during and after the reducing process, as a guide to the exposure to light.—An under-exposed plate developes slowly. By continuing the action of the Pyrogallic Acid, the high lights *become very black*, but the shadows are invisible, nothing but the yellow Iodide being seen on those portions of the plate. After treatment with the Cyanide, the picture shows well as a Positive, but by transmitted light all the minor details are invisible; the image is black and white, without any half-tone.

An over-exposed Negative developes rapidly at first, but soon begins to blacken slightly at every part of the plate. After the fixing is completed, nothing can often be seen by reflected light but a uniform grey surface of metallic Silver, without any appearance (or, at most, an indistinct one) of an image. By transmitted light the plate may appear of a red or brown colour, and the image is *faint* and dull. The clear parts of the Negative being obscured by the fogging, and the half-shadows having acted so long as nearly to overtake the lights, there is a want of proper *contrast;* hence the over-exposed plate is the exact converse of the under-exposed, where the contrast between lights and shadows is too well marked, from the absence of intermediate tints.

A Negative which has received the proper amount of exposure, usually possesses the following characters after the development is completed :—The image is partially but not fully seen by reflected light. In the case of a portrait, any dark portions of drapery show well as a Positive, but the features of the sitter are scarcely to be discerned. The plate has a general aspect as of fogging *about to commence,* but not actually established. By transmitted light the figure is bright, and appears to stand out from the glass: the dark shadows are clear, without any misty deposit of metallic Silver; the high lights black *almost* to complete opacity. The *colour* of the image however varies much with the state of the Bath and Collodion, and with the brightness of the light.

The remarks already made under the head of Positives, apply equally well to Negatives; that is, it will be difficult to secure gradation of tone, unless the object be *equally* illuminated, without any strong contrast of light and shade. Hence the direct rays of the sun are, as a rule, to be avoided, and curtains, etc., employed when practicable.

FIXING AND VARNISHING THE IMAGE.

After the development is completed, and the plate has been carefully washed by a stream of water, it may be brought out to the light and treated with the Hyposulphite or Cyanide, until the unaltered Iodide is entirely cleared off. Some use a Bath for the Cyanide; but it is doubtful whether much saving is effected by doing so. The plate is again to be carefully washed after the fixing;

Q

and especially if Hyposulphite of Soda be used. Three or four minutes in running water will not be too long, or the glass may be left in a dish of water for an hour or two. If such precautions are neglected, crystals form on drying, and the image is injured.

Collodion pictures should be protected by a coat of varnish, both Negatives and Positives having been known to fade when exposed to damp air without any covering (see p. 166). To prepare transparent varnish, Amber may be dissolved in Chloroform according to Dr. Diamond's formula;—about 80 grains of amber-beads or pipe-stems should be digested with one ounce of the Chloroform, and the clear portion separated by filtration. It may be poured on the plate in the same manner as Collodion, and dries up speedily into a hard and transparent layer. The Spirit Varnish ordinarily sold for Negatives requires the aid of heat to prevent the gum from chilling as it dries; the plate is first warmed *gently* and the varnish poured on and off in the usual way; it is then, whilst still dripping, held to the fire until the Spirit has evaporated. A few trials will render the operation easy to perform. White Lac dissolved in strong Alcohol or in Benzole has also been recommended for clear varnish.

Direct Positives are to be varnished, first with a layer of transparent varnish, and then with black japan. Suggett's patent jet is sometimes employed, but it has a disagreeable smell, and is apt to crack on drying. The best black japan used by coachmakers is more elastic and less liable to crack. Asphalt (4 oz.) dissolved in mineral Naphtha (10 oz.), with the addition of 30 grains of Caoutchouc dissolved in half an ounce of the same menstruum, is also said to stand well. A third formula contains black sealing-wax dissolved in Alcohol. In either case it will be best to apply first a layer of clear varnish to the film, and afterwards the black varnish, which should combine with the other without dissolving it.

Positives whitened with Bichloride of Mercury are injured by varnishing; they must therefore be backed up with black velvet, or Japan laid upon the opposite side of the glass. Many prefer taking the picture upon coloured glass, using only a layer of clear varnish; but in this case the Collodion side being left uppermost, the image is necessarily reversed.

SECTION II.

Directions for the use of Photographic Lenses.

Those who are comparatively unacquainted with the science of optics require simple rules to guide them in the choice of a photographic lens, and in the proper mode of using it.

Two kinds of Achromatic lenses are sold, the Portrait lens and the View lens; the former of which is constructed to admit a large volume of light, for the purpose of copying living objects, etc.

A convenient-sized Camera for small portraits is "the half-plate" with a lens of about $2\frac{1}{4}$ inches diameter, and giving a tolerably flat field on a surface of 5 inches by 4. Much however in this respect will depend upon the quality of the glass and also upon its focal length; a short focus lens taking a picture more quickly, but giving a smaller image, and a field which is misty towards the edge. There is also a great tendency to *distortion* of the image in portrait lenses of large aperture and short focus, such as those employed for operating in a dull light.

The "whole plate" portrait lens may be expected to cover $6\frac{1}{2}$ by $4\frac{3}{4}$ inches, and has a diameter of about $3\frac{1}{4}$ inches. It will take larger pictures than the last, but not necessarily in a shorter time; since, although the aperture for admitting the light is larger, the focal length is proportionately greater and the light less condensed.

The "quarter-plate" portrait lens of $1\frac{3}{4}$ inch diameter is useful for stereoscopic subjects and small portraits; which are usually more sharply defined when taken with a small lens.

The distance at which the Camera is to be placed from the sitter in taking a portrait, will depend upon the focal length of the lens. The effect of bringing the Camera nearer is to add to the size of the image, but at the same time to increase the chance of distortion; hence with every lens of full aperture, there is a practical limit to the size of picture which can be taken.

When it is required to obtain a large image with a small lens, a stop with a central aperture (which may be readily made of a piece of circular cardboard blackened with Indian ink) must be placed in front of the lens. This will diminish the amount of light, but will render the picture more distinct towards the edge, and bring a variety of objects at different distances into focus at the same time. With a stop attached, the lens may also be brought nearer to the object without distorting.

With regard to this subject of the distortion often produced by lenses, observe particularly, that with the portrait combination of full aperture, and especially when the powers of the glass are rather strained by its being advanced too near to the sitter,—all objects near to the lens will be *magnified*, and those more removed will appear diminished; hence, as the position of the sitter is never quite vertical, the Camera must be inclined a little *downwards*, or the hands and feet will be enlarged, the figure in fact becoming pyramidal with the base below; whereas on the other hand, if the inclination of the Camera be too great, the head and forehead will be enlarged, and the figure becomes a pyramid with the base above.

When groups are taken, arrange the objects as near as possible equidistant from the lens, and use a stop if practicable. Long-focus lenses are the best for this purpose, allowing the Photograph to be taken further off, and giving a greater variety of objects in focus at the same time.

Portrait lenses may often be advantageously substituted for View lenses in copying objects of still life which are *badly lighted*. The aperture of the lens being large, a

Negative can be obtained with an amount of light which would not suffice if a small stop were used.   On the other hand, if the light be unusually bright, the lens of full aperture is always the most likely, from its extent of reflecting surface, to produce a misty and indistinct image.   Hence the object should be well backed up with some neutral colour, or, if that cannot be done, a pasteboard funnel, projecting about a foot and a half, may be fastened in front of the lens, in order to exclude rays of light not immediately concerned in the formation of the image.   If the lens were turned towards distant objects brightly illuminated, and a portion of sky included, there would probably be diffused light, and consequent fogging of the plate on the application of the developer.   This effect will also invariably follow if the sun's rays be allowed to fall directly upon the glass.

Directions for finding the Plane at which the Sharpest Image can be obtained.—Non-Achromatic Lenses are understood by all to require correction for the chemical focus; but it is usually said of the compound glasses, that their two foci correspond.   The amateur is recommended, in order to avoid disappointment, to test the accuracy of this statement, and also to see that his Camera is constructed with care.   To do this, proceed as follows :—

First ascertain that the prepared sensitive plate falls precisely in the plane occupied by the ground glass.   Suspend a newspaper or a small engraving at the distance of about three feet from the Camera, and focus the letters occupying the centre of the field; then insert the slide, with a square of *ground glass* substituted for the ordinary plate (the rough surface of the glass looking inwards), and observe if the letters are still distinct.   In place of the ground glass, a transparent plate with a square of silver-paper which has been oiled or wetted, may be used, but the former is preferable.

If the result of this trial seems to show that the Camera is good, proceed to test the correctness of the Lens.—

Take a Positive Photograph with the full aperture of the portrait Lens, the central letters of the newspaper being carefully focussed as before. Then examine at what part of the plate the greatest amount of distinctness of outline is to be found. It will sometimes happen, that whereas the exact centre was focussed visually, the letters on a spot midway between the centre and edge are the sharpest in the Photograph. In that case the chemical focus is longer than the other, and by a distance equivalent to, but in the opposite direction of, the space which the ground glass has to be moved, in order to define those particular letters sharply to the eye.

When the chemical focus is the shorter of the two, the letters in the Photograph are indistinct at every portion of the plate; the experiment must therefore be repeated, the lens being shifted an eighth of an inch or less. Indeed it will be proper to take many Photographs at minute variations of focal distance before the capabilities of the lens will be fully shown.

The object of finding the point at which the sharpest image is obtained will also be assisted by placing several small figures in different planes and focussing those in the centre. This being done, if the more distant figures come out distinctly in the Photograph, the chemical focus is *longer* than the Visual, or *vice versâ* when the nearest ones are most sharply defined.

The Single Achromatic Lens.—A useful lens for landscape Photography is one of about 3 inches diameter and 15 inches focal length, which may be expected to cover a field of 10 inches by 8. With the lens, stops are supplied of various diameters, the largest of which will be useful in dull weather; the smaller when the field is required to be rendered sharp to the very edge.

The stop is arranged at a certain distance in front of the lens, and must not be moved. If it were brought close up to the glass, the field would not be so flat; the effect being then the same as that of a stop placed in front of a

Portrait Lens, viz. simply to cut off the outside portion of the glass.*

In taking Photographs of architectural and other subjects with vertical outlines, it is very important to have the Camera placed perfectly horizontal; since, if it be inclined either upwards or downwards, the perpendiculars will be destroyed and the object will appear of a pyramidal form, falling inwards or outwards, as before shown. It is convenient to rule the ground focussing glass with a number of parallel lines in both directions, which enables the operator at once to see that the position of the instrument is correct.

SECTION III.

Mode of copying Engravings, Etchings, etc.

The engraving to be Photographed should be removed from its frame (the glass causing irregular reflection) and suspended vertically and in a reversed position, in a good diffused light. A black cloth may be placed behind the picture with advantage if any surface likely to reflect light be presented to the lens.

The Camera must be fixed immovably, so as not to vibrate in the least degree when the cap of the lens is taken off. It should be pointed at right angles to the picture, and the focus determined in the ordinary way. Either a portrait or a single lens may be used, with a diaphragm sufficiently small to render the image distinct up to the edge.

It is not desirable to employ too thin a Collodion, since perfect opacity of the darkest parts of the Negative is essential. An old Collodion containing free Iodine is better than a contractile Collodion, as giving a more intense and clear image. Pure Collodion iodized with Iodide of Cadmium, if found wanting in intensity, may be at once ren-

* See this subject explained in ' Photographic Journal,' vol. ii. p. 133.

dered fit for use in copying engravings by adding Glycyr-rhizine (p. 209), until the dark parts of the negative become very opaque, and subsequently softening the excessive hardness, if necessary, by dropping alcoholic solution of Iodine into the Collodion until it reaches a straw-yellow tint. A second formula useful in iodizing Collodion for a similar purpose is as follows.

Iodide of Potassium . . . 4 grains.
Bromide of Potassium . . 1 grain.

This, with addition of Glycyrrhizine, will give a very black image.

Etchings, diagrams, and drawings with pencil or ink, without much middle-tint, if on thin paper, are easily co-pied without the aid of the Camera, by simply laying the sketch upon a sheet of Negative Paper, exposing for a brief time to the light, and developing with Gallic Acid. This yields a Negative which is employed for printing Positives in the usual way. Full directions on this subject will be found in the Second Section of the following Chapter.

A more simple plan, and one which will succeed when great delicacy is not required, consists in laying the sketch upon a sheet of Positive printing paper (a highly salted paper will be the best, as giving most intensity) and ex-posing to the light until a copy is obtained. All the de-tails are faithfully rendered in this way, but it is some-times difficult to obtain a Negative sufficiently black to yield a *vigorous* print.

SECTION IV.

Rules for taking Stereoscopic Photographs.

Binocular pictures of a large size, for the reflecting Ste-reoscope, may be taken with an ordinary View lens of about 15 inches focus. The ground glass of the Camera having been ruled with cross lines in the manner described at page 231, the position of some prominent object is marked upon one of the lines with a pencil, and the first view is taken. The stand is then moved laterally to the proper distance, and the Camera adjusted to its second position by shifting it until the marked object occupies the same place as before. The distance between the two positions should be about one foot when the foreground of the picture is twenty-five feet from the instrument, or four feet when it is at thirty or forty yards. But, as before shown at page 71, this rule is not to be followed implicitly, much depending upon the character of the picture and the effect desired.

Photographs for the lenticular Stereoscope are taken with small lenses of about $4\frac{1}{2}$ inches focus. For portraits, a Camera may advantageously be fitted with two double-combination lenses, of $1\frac{3}{4}$ inches diameter, exactly equal in focal length and in rapidity of action. The caps are removed simultaneously, and the pictures impressed at the same instant. The centres of the lenses may be separated by three inches when the Camera is placed at about six feet from the sitter, or four inches when the distance is increased to eight feet.

Pictures taken with a binocular Camera of this kind, require to be mounted in a reversed position to that which they occupy on the glass: for since the image of the Camera is *inverted,* when it is turned round and made erect, the right-hand picture will necessarily stand on the left side, and *vice versâ.*

Mr. Latimer Clark has devised an arrangement for taking stereoscopic pictures with a single Camera, which is exceedingly ingenious. Its most important feature is a contrivance for rapidly moving the Camera in a lateral direction without disturbing the position of the image upon the ground glass. This will be understood by a reference to the following woodcut.

"A strongly-framed Camera-stand carries a flat table, about 20 inches wide by 16, furnished with the usual ad-

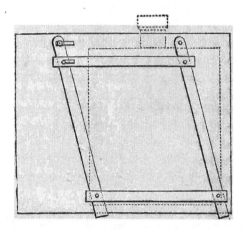

justments. Upon this are laid two flat bars of wood in the
direction of the object, and parallel, and about the width
of the Camera asunder. They are 18 inches in length;
their front ends carry stout pins, which descend into the
table and form centres upon which they turn. Their opposite ends also carry similar pins, but these are directed
upwards, and fit into two corresponding holes in the tail-
board of the Camera.

"Now when the Camera is placed upon these pins, and
moved to and fro laterally, the whole system exactly resembles the common parallel ruler. The two bars form
the guides, and the Camera, although capable of free lateral motion, always maintains a parallel position. In this
condition of things it is only suited to take stereoscopic
pictures of an object at an infinite distance; but to make
it move in an arc, *converging* on an object at any nearer
distance, it is only necessary to make the two guide-bars
approximate at their nearer end so as to converge slightly
towards the object; and by a few trials some degree of
convergence will be readily found at which the image will
remain as it were *fixed* on the focussing glass while the
Camera is moved to and fro. To admit of this adjustment,
one of the pins descends through a Slot in the table and

carries a clamping-screw, by means of which it is readily fixed in any required position.

"In order however to render the motion of the Camera smoother, it is advisable not to place it directly upon the two guides, but to interpose two thin slips of wood, lying across them at right angles, beneath the front and back of the Camera respectively (and which may be fixed to the Camera if preferred), and to dust the surfaces with powdered soap-stone or French chalk."

In addition to this arrangement for moving the Camera laterally, the *slide* for holding the sensitive plates must be modified from the common form. It is oblong in shape, and being about ten or eleven inches long, requires some little adaptation to fit it to the end of an ordinary Camera. The glasses are cut to about $6\frac{3}{4}$ inches by $3\frac{1}{4}$; and when coated with Iodide of Silver, the two images are impressed side by side, the plate being shifted laterally about $2\frac{1}{2}$ inches, at the same time and in the same direction as the Camera itself.

The operation of taking a portrait is thus performed. The focus having been adjusted for both positions, and the Camera and the slide both drawn to the left-hand, the door is raised and the plate exposed ; the Camera and the slide are then shifted to the right-hand, and the plate in its new position having been again exposed, the door is closed and the operation completed.*

Pictures taken with this instrument do not require to be reversed in mounting, the left picture being purposely formed on the right-hand side of the glass.

SECTION V.

On the Photographic delineation of Microscopic objects.

Many specimens of Micro-photography which have been exhibited are exceedingly elaborate and beautiful ; and

* See 'Photographic Journal,' vol. i. page 59.

their production is not difficult to one thoroughly acquainted with the use of the Microscope and with the
manipulations of the Collodion Process. It is important
however to possess a good apparatus, and to have it properly arranged.

The object-glass of the ordinary compound Microscope
is the only part actually required in Photography, but it
is useful to retain the *body* for the sake of the adjustments,
and the mirrors used in the illumination. The *eye-piece*
however, which simply magnifies the image formed by the
object-glass, is not necessary, since the same effect of enlargement may be obtained by lengthening out the dark
chamber, and throwing the image further off.

Arrangement of the Apparatus. — The Microscope is
placed with its body in a horizontal position, and the eye-
piece being removed, a tube of paper, properly blackened
in the interior, or lined with black velvet, is inserted into
the instrument, to prevent irregular reflection of light from
the sides.

A dark chamber of about two feet in length, having at
one end an aperture for the insertion of the eye-piece end
of the body, and at the other a groove for carrying the
slide containing the sensitive plate, is then attached; care
being taken to stop all crevices likely to admit diffused
light. An ordinary Camera may be employed as the dark
chamber, the lens being removed, and the body lengthened
out if required by a conical tube of gutta-percha, made to
fasten into the flange of the lens in front. The whole apparatus should be placed exactly in a straight line, that the
ground glass used in focussing may fall at right angles to
the axis of the Microscope.

The length of the chamber, measuring from the object-
glass, may be from two to three feet, according to the size
of image required; but if extended beyond this, the pencil
of light transmitted by the object-glass is diffused over too
arge a surface, and a faint and unsatisfactory picture is
the result. The object should be illuminated by sunlight

if it can be obtained, but a bright diffused daylight will succeed with low-power glasses, and especially when Positives are taken. Employ the *concave* mirror for reflecting the light on the object in the latter case; but in the former the *plane mirror* is the best, except with powers exceeding a quarter of an inch, and of large angular aperture.

The image upon the ground glass should appear bright and distinct, and the field of a circular form and evenly illuminated; when this is the case, all is ready for inserting the sensitive plate.

The time of exposure must be varied according to the intensity of the light, the sensibility of the Collodion, and the degree of magnifying power; a few seconds to a minute will be about the extremes; but minute directions are not required, as the operator, if a good Photographer, will easily ascertain the proper time for exposing (see page 224).

At this point a difficulty will probably occur from the plane of the chemical focus not corresponding, as a rule, with that of the visual focus. This arises from the fact that the object-glasses of Microscopes are "over-corrected" for colour, in order to compensate for a little chromatic aberration in the eye-piece. The violet rays, in consequence of the over-correction, are projected *beyond* the yellow, and hence the focus of chemical action is further from the glass than the visible image.

The allowance may be made by shifting the sensitive plate, or, what amounts to the same thing, by removing the object-glass a little *away* from the object with the fine adjustment screw; the latter is the most convenient. The exact distance must be determined by careful experiment for each glass; but it is greatest with the low powers, and decreases as they ascend.

Mr. Shadbolt gives the following as a guide :—" An inch and a half objective of Smith and Beck's make required to be shifted 1-50th of an inch, or two turns of *their* fine adjustment; a 2-3rds of an inch, 1-200th of an inch, or half a turn; and a 4-10ths of an inch, 1-1000th of an inch, or about two divisions of the adjustment. With the 1-4th and higher powers, the difference between the foci was so small as to be practically unimportant."

There is also reason to think that the *kind of light* employed has an influence upon the separation of the foci. Mr. Delves finds that with sunlight the difference between them is very small even with the low powers, and inappreciable with the higher; whereas in using diffused daylight which has undergone a previous reflection from white clouds, it is considerable.

The object-glasses of the same maker, and particularly those of different makers, also vary much; so that it will be necessary to test each glass separately, and to register the allowance which is required.

Having found the chemical focus, the principal difficulty has been overcome, and the remaining steps are the same in every respect as for ordinary Collodion Photographs.

To those who cannot devote their time to Photography during the day, Mr. Shadbolt's observations on the use of artificial light may be of service. He employs *Camphine*, which gives a whiter flame than gas, or a moderator lamp; placing the source of light in the focus of a plano-convex lens of $2\frac{1}{2}$ to 3 inches diameter (the flat side towards the lamp), and condensing the parallel rays so obtained on the object, by a second lens of about $1\frac{1}{2}$-inch diameter and 3-inch focus.

This mode of illumination, being feeble in chemical rays, is best adapted for object-glasses of low power. The exposure required to produce a Negative impression with the one-inch glass may be from three to five minutes. As the sensitive plate would be liable to become dry during that time, it is recommended to coat it with some preservative solution by the modes described in the sixth Chapter. Mr. Crookes having lately shown that the Bromide of Silver is more sensitive than the Iodide to artificial light, a mixture of the two salts may conveniently be used (see pp. 66 and 232).

The development may be conducted in the same manner as that for preserved sensitive plates; fixing with Cyanide of Potassium before the development is fully complete, if any tendency to fogging is observed (see page 224).

The Rev. W. Towler Kingsley has communicated a process by which very beautiful Microscopic Photographs have been obtained. He illuminates (in the absence of sunlight) with the brilliant light produced by throwing a jet of mixed Oxygen and Hydrogen gases upon a small cone of Lime or Magnesia. Particular stress is laid upon the object-glass of the Microscope being a good one for the purpose; and indeed all who have given attention to the subject are agreed upon this point—that there is a considerable difference in the Photographic value of objectives, and this independent of the angular aperture of the glass.

CHAPTER IV.

THE PRACTICAL DETAILS OF PHOTOGRAPHIC
PRINTING.

THIS Chapter is divided as follows :—

SECTION I.—The ordinary direct process of positive
printing.

SECTION II.—Positive printing by development.

SECTION III.—The mode of toning Positives by Sel d'or.

SECTION IV.—On printing enlarged or reduced Positives, transparencies, etc.

SECTION I.

Positive Printing by the direct action of Light.

This includes—the preparation of sensitive paper,—of
fixing and toning Baths,—and the manipulatory details of
the process.

Selection of Paper for Photographic Printing.—The
ordinary varieties of paper sold in commerce are not well
adapted for the production of Positive prints. Papers are
manufactured purposely which are more smooth and uniform in texture. Many samples of even the finest paper
are however defective, and hence each sheet should be examined separately by holding it against the light, and if
spots or irregularities of texture are seen, it should be rejected. These spots usually consist of small particles of
brass or iron, which, when the paper is rendered sensitive,

decompose the Nitrate of Silver and leave a circular mark very noticeable after fixing.

The foreign papers, French and German, are different from the English. They are porous and sized with starch, the English being sized with gelatinous animal matter. In all cases there is a difference in smoothness between the two sides of the paper, which may be detected by holding each sheet in such a manner that the light strikes it at an angle; the wrong side is that on which dark wavy bands, of an inch to an inch and a half in breadth, are seen, caused by the strips of felt on which the paper was dried. With most qualities of paper no difficulty whatever will be experienced in detecting the broad and regular bands above referred to; but when they cannot be seen, the wrong side of the sheet may be known by wire markings crossing each other, or if the paper be wetted at the corner, one side may appear evidently smoother than the other.

PREPARATION OF SENSITIVE PAPER.

There are three principal varieties of sensitive paper in common use, viz. the Albuminized, the plain, and the Ammonio-Nitrate paper.

Formula I. Preparation of Albuminized Paper.—This includes the salting and albuminizing, and the sensitizing with Nitrate of Silver.

The Salting and Albuminizing.—Take of

Chloride of Ammonium, or Pure
   Chloride of Sodium . . . . 200 grains.
Water . . . . 10 fluid ounces.
Albumen . . . . 10 fluid ounces.

If distilled water cannot be procured, rain water or even common spring water* will answer the purpose. To obtain the Albumen, use new-laid eggs, and be careful

* If the water contained much Sulphate of Lime, it is likely that the sensitiveness of the paper would be impaired (?).

R

that in opening the shell the yolk is not broken; each egg will yield about one fluid ounce of Albumen.

When the ingredients are mixed, take a bundle of quills or a fork, and beat the whole into a perfect froth. As the froth forms, it is to be skimmed off and placed in a flat dish to subside. The success of the operation depends entirely upon the manner in which this part of the process is conducted;—if the Albumen be not thoroughly beaten, flakes of animal membrane will be left in the liquid, and will cause streaks upon the paper. When the froth has partially subsided, transfer it to a tall and narrow jar, and allow to stand for several hours, that the membranous shreds may settle to the bottom. Then pour off the upper clear portion, which is fit for use. Albuminous liquids are too glutinous to run well through a paper filter, and are better cleared by subsidence.

A more simple plan than the above, and one equally efficacious, is to fill a bottle to about three parts with the salted mixture of Albumen and water, and to shake it well for ten minutes or a quarter of an hour until it loses its glutinosity and can be poured out smoothly from the neck of the bottle. It is then to be transferred to an open jar, and allowed to settle as before.

The solution prepared by the above directions will contain exactly ten grains of salt to the ounce, dissolved in an equal bulk of Albumen and water. Some operators employ the Albumen alone without an addition of water; but this commonly gives a highly varnished appearance, which is thought by most to be objectionable. Much however will depend upon the kind of paper which is employed, certain varieties taking more gloss than others; Papier Rive, for instance, often requires the Albumen to be nearly or quite undiluted.

The principal difficulty in Albuminizing paper, is to avoid the occurrence of *streaky lines*, which, when the paper is rendered sensitive, *bronze* strongly under the influence of the light. To avoid them, use the eggs quite fresh, and lower the paper on to the liquid by one steady movement; if a pause be made, a line will probably be formed. Some papers are not readily wetted by the Albumen, and when such is the case, a few drops of spirituous solution of bile, or a fragment of the prepared Ox-Gall sold by the artists'-colourmen, will be found a useful adjunct. Care must be taken however not to add an excess, or the Albumen will be rendered too fluid, and will sink into the paper, leaving no gloss.

In salting and albuminizing Photographic paper by the formula above given, it is found that each quarter-sheet, measuring eleven inches by nine inches, removes one fluid drachm and a half from the bath, equivalent to about one grain and three-quarters of salt (including droppings). In salting plain paper, each quarter-sheet takes up only one drachm; so that the glutinous nature of the Albumen causes a third part more of salt to be retained by the paper.

English papers are not good for albuminizing; they do not take the Albumen properly, and curl up when laid upon the liquid: the process of toning the prints is also slow and tedious. The thin negative paper of Canson, the Papier Rive, and Papier Saxe, have succeeded with the writer better than Canson's Positive paper, which is often recommended; they have a finer texture, and give more smoothness of grain.

To apply the Albumen, pour a portion of the solution into a flat dish to the depth of half an inch. Then, having previously cut the paper to the proper size, take a sheet by the two corners, bend it into a curved form, convexity downwards, and lay it upon the Albumen, the centre part first touching the liquid, and the corners being lowered gradually. In this way all bubbles of air will be pushed forwards and excluded. One side only of the paper is wetted: the other remains dry. Allow the sheet to rest upon the solution for *one minute and a half*, and then raise it off, and pin it up by two corners. If any circular spots, free from Albumen, are seen, caused by bubbles of air, replace the sheet for the same length of time as at first.

The paper must not be allowed to remain upon the salting Bath much longer than the time specified, because the solution of Albumen being *alkaline* (as is shown by the strong smell of Ammonia evolved on the addition of the Chloride of Ammonium) tends to remove the size from the paper and to sink in too deeply; thus losing its surface gloss.

Albuminized paper will keep a long time in a dry place. Some have recommended to press it with a heated iron, in order to coagulate the layer of Albumen upon the surface; but this precaution is unnecessary, since the coagulation is perfectly effected by the Nitrate of Silver used in the sensitizing; and it is doubtful whether a layer of *dry* Albumen would admit of coagulation by the simple application of a heated iron.

To render the paper sensitive.—This operation must be conducted by the light of a candle, or by yellow light. Take of

> Fused Nitrate of Silver . . . 60 grains.
> Glacial Acetic Acid . . . . . $\frac{1}{3}$ minim.
> Distilled Water . . . . . . 1 ounce.

Prepare a sufficient quantity of this solution, and lay the sheet upon it in the same manner as before. Three minutes' contact will be sufficient with the thin Negative paper, but if the Canson Positive paper be used, four or five minutes must be allowed for the decomposition. The papers are raised from the solution by a pair of bone forceps or common tweezers tipped with sealing-wax; or a pin may be used to lift up the corner, which is then taken by the finger and thumb and allowed *to drain a little* before again putting in the pin, otherwise a white mark will be produced upon the paper, from decomposition of the Nitrate of Silver. When the sheet is hung up, a small strip of blotting-paper suspended from the lower edge of the paper will serve to drain off the last drop of liquid.

A Bath prepared by the above formula is stronger than is really necessary. Forty grains of Nitrate to the ounce of water is abundantly sufficient if the sample be pure; but it must be borne in mind that the *strength* of the Bath diminishes *rapidly* by use, and hence, when the prints begin to be wanting in vigour, with pale shadows and perhaps a spotted appearance, an addition of Nitrate of Silver must be made. Fused Nitrate of Silver is recommended in preference to the crystallized Nitrate, on account of the latter being occasionally contaminated with an impurity alluded to at page 101. This when present will be likely to redden the pictures and to interfere with the rapidity of bronzing.

The solution of Nitrate of Silver becomes after a time discoloured by the Albumen, but may be used for sensitizing until it is nearly black. The colour can be removed by Animal Charcoal,* but a better plan is to use the "kaolin," or pure white china clay. This substance often contains Carbonate of Lime, and effervesces with acids: it must in such a case be purified by washing in vinegar, or the Bath will become alkaline, and dissolve off the Albumen. It has been stated that an addition of Alcohol to the Nitrate Bath prevents it discolouring with Albumen.

Sensitive albuminized paper will usually keep for several days, if protected from the light, but afterwards turns yellow from partial decomposition.

Formula II. *Preparation of plain paper.*—Take of

Chloride of Ammonium or Sodium . 160 grains.
Purified Gelatine . . . . 20 grains.
Iceland Moss† . . . . 60 grains.
Water . . . . 20 ounces.

* Common Animal Charcoal contains Carbonate and Phosphate of Lime, the former of which renders the Nitrate of Silver *alkaline;* purified Animal Charcoal is usually acid from Hydrochloric Acid.

† Iceland Moss is recommended because the writer finds that Positives

Pour boiling water upon the Moss and Gelatine and stir until the latter is dissolved, then cover the vessel and set aside until cold ; add the salt, and strain.

Use Papier Saxe or Towgood's paper,* floated upon the salting Bath in the same manner as directed for Albumen at p. 243.

Render sensitive by floating for two or three minutes upon a solution of Nitrate of Silver, 40 grains to the ounce. Thirty grains to the ounce, or less, will be sufficient if the sample be pure ; but in that case occasional additions of fresh Nitrate of Silver must be made, as the Bath loses strength.

A second Formula for plain paper.—Take of

Chloride of Ammonium	200 grains.
Citrate of Soda†	200 „
Gelatine	20 „
Water	20 fluid ounces.

If Towgood's or any English paper be used, the Citric Acid, Carbonate of Soda, and Gelatine may be omitted. With a foreign paper the Citrate tends to give a purple tone to the Positive, when toned by Sel d'or, but the gold toning Bath must be in active order, or the prints will be too red. The Citric Acid also should not be in excess over the alkaline Carbonate.

Render sensitive by floating for three minutes upon a Nitrate Bath of sixty grains to the ounce of water.

Formula III. *Ammonio-Nitrate Paper.*—This is always prepared without Albumen, which is dissolved by Ammonio-Nitrate of Silver. Take of

so printed stand the action of destructive tests better than prints on plain paper, and equal to prints upon Ammonio-Nitrate paper.

* The writer does not recommend the Positive paper of De Canson, having noticed that prints upon that paper do not withstand the action of sulphuretting agents so well as others (?).

† This salt may be obtained at the operative chemists; or it may be prepared extemporaneously by neutralizing 112 grains of pure Citric Acid, free from Tartaric Acid, with 133 grains of the dried Bicarbonate or "Sesquicarbonate" of Soda, used for effervescing draughts.

Chloride of Ammonium . . . . . . 100 grains.
Citrate of Soda . . . . 200 „
Gelatine . . . . 20 „
Water . . . . 20 fluid ounces.

Dissolve the Gelatine by the aid of heat; add the other ingredients, and filter. The solution cannot be kept longer than two or three weeks without becoming mouldy. The Saxony paper, or Towgood's English paper, may be employed; the Gelatine and Citrate being retained or omitted, according to the taste of the operator and the mode of toning which is adopted.

Render sensitive by a solution of Ammonio-Nitrate of Silver, 60 grains to the ounce of water, which is prepared as follows :—

Dissolve the Nitrate of Silver in one-half of the total quantity of water. Then take a pure solution of Ammonia and drop it in carefully, stirring meanwhile with a glass rod. A brown precipitate of Oxide of Silver first forms, but on the addition of more Ammonia it is redissolved.* When the liquid appears to be clearing up, add the Ammonia very cautiously, so as not to incur an excess. In order still further to secure the absence of free Ammonia, it is usual to direct, that when the liquid becomes perfectly clear, a drop or two of solution of Nitrate of Silver should be added until a *slight turbidity* is again produced. Lastly, dilute with water to the proper bulk. If the crystals of Nitrate of Silver employed contain a large excess of free Nitric Acid, no precipitate will be formed on the first addition of Ammonia. The free Nitric Acid, producing *Nitrate of Ammonia* with the alkali, keeps the Oxide of Silver in solution. This cause of error however is not likely to happen frequently, since the amount of Nitrate of Ammonia required to prevent all precipitation would be considerable. From the same reason,

* If the excess of Ammonia does not readily dissolve it, probably the Nitrate of Silver is impure.

viz. the presence of Nitrate of Ammonia, it is often useless to attempt to convert an old Nitrate Bath already used for sensitizing, into Ammonio-Nitrate.

Ammonio-Nitrate of Silver should be kept in a dark place, being more prone to reduction than the Nitrate of Silver.

Sensitizing paper with Ammonio-Nitrate.—It is not usual to *float* the paper when the Ammonio-Nitrate of Silver is used. If a bath of this liquid were employed, it would not only become quickly discoloured by the action of organic matter dissolved out of the papers, but would soon contain abundance of free Ammonia (see the Vocabulary, Part III., art. "Ammonio-Nitrate"); and an excess of Ammonia in the liquid produces an injurious effect by dissolving away the sensitive Chloride of Silver.

The Ammonio-Nitrate is therefore applied with a glass rod, or by brushing, and in neither case is any of the liquid which has once touched the paper allowed to return into the bottle.

Brushes are manufactured purposely for applying Silver solutions, but the hair is soon destroyed unless the brush be kept scrupulously clean. Lay the salted sheet upon blotting-paper, and wet it thoroughly by drawing the brush first lengthways and then across. Allow it to remain flat for a minute or so, in order that a sufficient quantity of the solution may be absorbed (you will see when it is evenly wet by looking along the surface), and then pin up by the corner in the usual way. If, on drying, *white lines* appear at the points last touched by the brush, it is probable that the Ammonio-Nitrate contains free Ammonia.

The employment of a glass rod is a very simple and economical mode of applying Silver solutions. Procure a flat piece of board somewhat smaller than the sheet to be operated on, and having turned over the edges of the paper, secure them with a pin. Next bring the board near to the corner of the table, and laying the glass rod along the edge of the paper, allow the fluid to drop into the groove so formed; then carry the rod directly across the sheet, when an even wave of fluid will be spread over the surface. A pipette made of glass tubing, when dipped into the bottle and the upper end closed with the finger, will withdraw as much of the Ammonio-Nitrate as is required; and if a scratch be made upon the tube at a point corresponding to 30 or 40 minims, it will be found sufficient for a quarter sheet of the Papier Saxe.

Ammonio-Nitrate paper, however prepared, cannot be kept many hours without becoming brown and discoloured.

Use of a solution of Oxide of Silver in Nitrate of Ammonia.—The great objection to the use of Ammonio-Nitrate of Silver is the *decomposition* which it sometimes experiences by keeping, metallic Silver separating and Ammonia being set free. To obviate this liberation of Ammonia, the Author employs *Nitrate of Ammonia* as the solvent for the Oxide of Silver. The solution is prepared as follows:—Dissolve 60 grains of Nitrate of Silver in half an ounce of water, and drop in Ammonia until the precipitated Oxide of Silver is exactly re-dissolved. Then divide this solution of Ammonio-Nitrate of Silver into two equal parts, to *one* of which add Nitric Acid cautiously, until a piece of immersed litmus-paper is reddened by an excess of the acid; then mix the two together, fill up to one ounce with water, and filter from the milky deposit of Chloride or Carbonate of Silver, if any be formed.

This solution of Oxide of Silver in Nitrate of Ammonia appears to possess all the advantages of the Ammonio-Nitrate without the inconvenience of liberating so much free Ammonia upon the surface of the sensitive sheets.

Hints in selecting from the above Formulæ.—Albuminized paper is the most simple and generally useful; it is well fitted for small portraits and stereoscopic Photographs. The Ammonio-Nitrate Process requires more experience, but gives excellent results when black tones are required: it may be used for larger portraits, engravings, etc.

Plain paper rendered sensitive by floating upon a Bath of Nitrate of Silver is easier of manipulation than the Ammonio-Nitrate, and will be found to be better adapted for toning by the Sel d'or Bath (p. 267) than the Albuminized Paper.

PREPARATION OF THE FIXING AND TONING BATH.

Take of

Chloride of Gold	4 grains.
Nitrate of Silver	16 grains.
Hyposulphite of Soda*	4 ounces.
Water	8 fluid ounces.

Dissolve the Hyposulphite of Soda in four ounces of the water, the Chloride of Gold in three ounces, the Nitrate of Silver in the remaining ounce; then pour the diluted Chloride by degrees into the Hyposulphite, stirring with a glass rod; and afterwards the Nitrate of Silver in the same way. This order of mixing the solutions is to be strictly observed: if it were reversed, the Hyposulphite of Soda being added to the Chloride of Gold, the result would be the reduction of Metallic Gold; Hyposulphite of Gold, which is formed, being an unstable substance, and not capable of existing in contact with unaltered Chloride of Gold. If however it be dissolved by Hyposulphite of Soda *immediately* on its formation, it is rendered more permanent, by conversion into a double salt of Soda and Gold.

In place of Nitrate of Silver, recommended in the formula, *Chloride of Silver* may be used, but not *Iodide of Silver*, as the formation of Iodide of Sodium would be objectionable (p. 136). For the same reason it is better not to add any part of the Hyposulphite Bath used for fixing Negatives, to the Positive colouring solution.

* The common kind of Hyposulphite of Soda occurring in yellow and discoloured masses, is too impure for use in Photography, and requires re-crystallization.

. This toning Bath is not to be employed immediately after mixing, but should be set aside until a portion of *Sulphur* (produced by free Hydrochloric Acid, and Tetra-thionate of Soda reacting upon the Hyposulphite) has subsided. It will be very active at the expiration of a few days or a week; but upon keeping for a longer time, loses much of its efficacy by a process of spontaneous change.

The immersion of prints also lessens the quantity of Gold; and hence, when the Bath begins to work slowly, more of the Chloride must be added, the Sulphur being allowed to deposit as before. Filtration through blotting-paper will not be required.

The writer finds that after a certain time, when the Bath has been long used, and organic matters, Albumen, etc., have accumulated in it, it is better, and more economical, to throw away what remains, and to prepare a new solution. The addition of Chloride of Gold to an old Bath will not always make it work as quickly as one recently mixed.

THE MANIPULATORY DETAILS OF PHOTOGRAPHIC PRINTING.

These include—the exposure to light, or printing properly so called; the fixing and toning; and the washing, drying, and mounting of the proof.

The Exposure to Light.—For this purpose reversing frames are sold, which admit of being opened at the back, in order to examine the progress of the darkening by light, without producing any disturbance of position.

Simple squares of glass however succeed equally well, when a little experience has been acquired. They may be held together by the wooden clips sold at the American warehouses at one shilling per dozen. The lower plate should be covered with black cloth or velvet.

Supposing the frame to be employed, the shutter at the back is removed, and the Negative laid flat upon the glass, Collodion side uppermost. A sheet of sensitive paper is then placed upon the Negative, sensitive side downwards, and the whole tightly compressed by replacing and bolting down the shutter.

This operation may be conducted in the dark room; but unless the light be strong, such a precaution will not be required. The time of exposure to light varies much with the density of the Negative and the power of the actinic rays, as influenced by the season of the year and other obvious considerations. As a general rule, the best Negatives print slowly; whereas Negatives which have been under-exposed and under-developed print more quickly.

In the early spring or summer, when the light is powerful, probably about ten to fifteen minutes will be required; but from three-quarters of an hour to an hour and a half may be allowed in the winter months, even in the direct rays of the sun.

It is always easy to judge of the length of time which will be sufficient, by exposing a small slip of the sensitive paper, *unshielded*, to the sun's rays, and observing how long it takes to reach the *coppery stage* of reduction. Whatever that time may be, nearly the same will be occupied in the printing, if the Negative be a good one.

When the darkening of the paper appears to have proceeded to a considerable extent, the frame is to be taken in and the picture examined. If squares of plate glass are used to keep the Negative and sensitive paper in contact, some difficulty may be experienced at first in returning it precisely to its former position after the examination is complete, but this will easily be overcome by practice. The finger and thumb should be fixed on the lower corners or edge, and the plate raised evenly and quickly.

If the exposure to light has been sufficiently long, the print appears *slightly darker* than it is intended to remain. The toning Bath dissolves away the lighter shades, and reduces the intensity, for which allowance is made in the exposure to light. A little experience soon teaches what is the proper point; but much will depend upon the state of the toning Bath; and albuminized paper will require to be printed somewhat more deeply than plain paper.

If, on removal from the printing-frame, a peculiar *spotted* appearance is seen, produced by unequal darkening of the Chloride of Silver, either the Nitrate Bath is too weak, the sheet removed from its surface too speedily, or the paper is of inferior quality.

On the other hand, if the general aspect of the print is a rich chocolate-brown in the case of Albumen, a dark slate-blue with Ammonio-Nitrate Paper, or a reddish purple with paper prepared with Chloride and Citrate of Silver, probably the subsequent parts of the process will proceed well.

If, in the exposure to light, the shadows of the proof become very decidedly *coppery* before the lights are sufficiently printed, the Negative is in fault. Ammonio-Nitrate paper highly salted is particularly liable to this fault of excess of reduction, and especially so if the light be powerful; hence it is best, in the summer months, not to print by the direct rays of the Sun. This point is important also, because the excessive heat of the Sun's rays often cracks the glasses by unequal expansion, and glues the Negative firmly down to the sensitive paper. An exception however may be made in the case of Negatives of great intensity; which are printed most successfully upon a weakly sensitized paper (p. 124) exposed to the full rays of the Sun; a feeble light not fully penetrating the dark parts.

The fixing and toning of the proof.—No injury results from postponing this part of the process for many hours, provided the print be kept in a dark place.

The mode often followed is to immerse the Positive in the Hyposulphite Bath in the state in which it comes from the printing-frame; moving it about in the liquid in order to displace air-bubbles, which, if allowed to remain, produce spots. But the Author, for reasons given in the first part of the Work (pp. 129 and 165), recommends that the print should first be washed in common water until the soluble Nitrate of Silver has been removed.* This is known to be the case when the liquid flows away clear; the first milkiness being caused by the soluble Carbonates and Chlorides in the water precipitating the Nitrate of Silver. Greater security is thus afforded that the print will be toned in a really permanent manner, since after removing the Nitrate of Silver from the proof, the Bath does not work quickly unless the supply of Gold be well maintained.

Immediately on coming in contact with the Hyposulphite of Soda in the fixing and toning Bath, the chocolate brown or violet tint of the Positive disappears, and leaves the image of a red tone. Albumen proofs become brick red; Ammonio-Nitrate a sepia or brown-black. If the colour is unusually *pale* at this stage, probably the Silver Bath is too weak, or the quantity of Chloride of Ammonium or Sodium insufficient.

After the print has been thoroughly reddened, the *toning* action begins, and must be continued until the desired effect is obtained. This may happen in from ten minutes to a quarter of an hour, if the solution is in good working order and the thermometer at 60°; but much depends upon the temperature, and the activity of the Bath. English papers, and especially the same prepared with Albumen, tone more slowly than foreign papers plain salted.

The brown and purple tints are an earlier stage of coloration than the black tones, and therefore the latter require more time. It must be borne in mind however that prolonged immersion in the Bath is favourable to sulphuration and yellowness; tending also to render the image unstable and liable to fade in the half-tones. This fading may not be seen decidedly whilst the print is in the Bath, but will show itself in the after-processes of washing and drying.

The ultimate colour of the Print will vary much with the density of the Negative and the character of the sub-

* This water must be free from Hyposulphite of Soda, or the print will become discoloured.

ject; copies of line engravings, having but little half-tone, are easily obtained of a dark shade resembling the original impression.

Some advise that on removal from the toning Bath the Print should be soaked in new Hyposulphite for ten minutes, to complete the fixation; but this precaution is not required with a Bath of the strength given in the formula. An analysis of an old Bath which had been extensively used, indicated only ten grains of Hyposulphite of Silver to the ounce, so that it was far from saturated.

The occasional addition of fresh crystals of Hyposulphite of Soda to keep up the strength of the Bath, is useful, the exact quantity added not being material.

The washing, drying, and mounting of the Positive Proofs. —It is essential to wash out every trace of Hyposulphite of Soda from the Print if it is to be preserved from fading, and to do this properly requires considerable care.

Always wash with *running water* when it can be obtained, and choose a large shallow vessel exposing a considerable surface in preference to one of lesser diameter. A constant dribbling of water must be maintained for four or five hours, and the prints should not lie together too closely, or the water does not find its way between them. (see the remarks at p. 162).

When running water cannot be obtained, proceed as follows:—first wash the Prints gently, to remove the greater part of the Hyposulphite solution. Then transfer them to a large shallow pan, in which may be placed as many Prints as it will conveniently hold. Leave them in for about a quarter of an hour, with occasional movement, and then *pour off the water quite dry.* This point is important, viz. to drain off the last portion of liquid completely before adding fresh water. Repeat the process of changing at least five or six times, or more, according to the bulk of water, number of Prints, and degree of attention paid to them.

Lastly, proceed to remove the size from the Print by

immersion in boiling water.*　This process will give some idea of the permanency of the tints, since, if they become dull and red, *and do not darken on drying*, the Print is probably toned with ut Gold. Ammonio-Nitrate and plain paper Prints prepared on foreign papers by the modes described in this Work, may be expected to stand the test of boiling water; Albumen Prints and Positives on English paper are a little reddened, although not to an objectionable degree.

The size may also be effectually removed from the Print by the common Carbonate of Soda used in washing, although the former process is recommended as the most secure.　Dissolve about a handful of the Soda in a pint of water, and when the milky deposit, if any occurs, has subsided, immerse the washed Positives for twenty minutes or half an hour. The Soda renders the paper quite porous, but produces no alteration of tint.　If the process be properly performed, ink will *run* in attempting to write upon the back of the finished picture.　After removal from the Soda Bath a second washing will be required, but the time of the first washing may be proportionally shortened. Here a difficulty will occur with many kinds of water; the Carbonate of Soda precipitating *Carbonate of Lime*, in the form of a white powder which obscures the picture. To obviate this, use *rain water* until the greater part of the alkaline salt has been removed, and do not allow a stationary layer of liquid to rest too long upon the Print. The New River water supplied to many parts of London, being comparatively soft, answers perfectly, and produces no white deposit, if the proofs are moved about occasionally.

When the Prints have been thoroughly washed, blot them off between sheets of porous paper and hang up to

---

* The Print must be well washed in cold water, to remove the Hyposulphite, before using the hot water; or the half-tones will be liable to be darkened, or changed to incipient yellowness, by sulphuration. This point is important as regards the permanency.

dry. Some press them with a hot iron, which darkens the colour slightly, but does so in an injurious manner when Hyposulphite of Soda is left in the paper.

Albumen proofs when dry are sufficiently bright without further treatment; but in the case of plain paper, salted simply, the effect is improved by laying the Print face downwards upon a square of plate-glass and rubbing the back with an agate burnisher, sold at the artists' colour-men's. This hardens the grain of the paper and brings out the details of the picture. Hot-pressing has a similar effect and is often employed.

Mount the proofs with a solution of Gelatine in hot water, freshly made; the best Scotch glue answers well. Gum water, prepared from the finest commercial gum, and free from acidity, may also be used, but it should be made very thick, that it may not sink into the paper, nor produce an unpleasant "cockling up" of the cardboard, which is caused by the damp and expanded print *contracting* as it dries.

Caoutchouc dissolved in mineral Naphtha to the consistence of thick glue or gold-beaters' size, is employed by many for mounting Photographic Prints; it may be obtained at the varnish shops, and is sold in tin boxes. The mode of using it is as follows:—with a broad brush made of stiff bristles, apply the cement to the back of the picture; then take a strip of glass with a straight edge, and by drawing it across the paper, scrape off as much as possible of the excess. The print will then be found to adhere very readily to the cardboard, without causing expansion or cockling; and any portion of the cement which oozes out during the pressing may, when dry, be removed with a penknife without leaving a stain.

REMARKS UPON THE WANT OF CORRESPONDENCE BETWEEN THE FORMULÆ OF DIFFERENT OPERATORS.

The formulæ for Positive printing given in the works

s

on practical Photography exhibit great variety; and it has been proposed to attempt to reduce them to more uniform proportions. This cannot however easily be done, both on account of the difference in the structure and preparation of the various Photographic papers, and also because the mode of applying the solutions is not always the same.

Take as an illustration the following process, which has long been recommended for its simplicity, and which is in every respect a good one :—Dissolve 40 grains of Chloride of Ammonium in 20 ounces of Distilled Water, and *immerse* about a dozen sheets of Towgood's Positive paper, removing air-bubbles with a camels'-hair brush. When the last sheet has been placed in the liquid, turn the batch over and take them out one by one, so that each sheet, remaining in the liquid at least ten minutes, may be thoroughly saturated. When dry, excite by brushing with a 40 or 60-grain solution of Ammonio-Nitrate of Silver in the usual way.

Now this formula contains less than one-fifth of the amount of salt often employed, and if a thick foreign paper sized with starch, such as Canson's Positive, were *floated* upon such a salting Bath, it would be difficult to obtain a good picture. By *immersing* however a paper sized with Gelatine like the one recommended, a much larger quantity of salt is retained upon the surface, and the film is sufficiently sensitive. There are three modes of applying solutions, viz. by brushing, floating, and immersion. The quantity of solution left on the paper varies with each, and consequently each requires a different formula. Immersion in a strong salting Bath tends to give a coarse picture wanting in definition; whereas the plan of brushing a weak salting solution, produces a paper deficient in sensitiveness, and yielding a pale red image without proper depth of shadow.

But independent of these differences, the chemical nature of the *size* employed also influences the toning of the

Print. For instance, in the process above given, if the Positives, after having been fully toned in the Gold Bath, and washed in cold water, be treated with *boiling* water, the tint immediately changes to a dull red; but on blotting off between sheets of bibulous paper and pressing with a hot iron, the dark tones are restored.

This destruction of the tint by boiling water, and its restoration by *dry heat,* is due in great part to the animal substance employed in sizing the paper; and it will be found that prints upon a foreign paper, such as the Saxony Positive, salted with a ten-grain solution and sensitized with Ammonio-Nitrate, do not lose their tones in hot water and are not much darkened by ironing.

The peculiarity of the sizing of the English Photographic papers must therefore be borne in mind, and allowance made for the additional sensitiveness and alteration of colour which it produces. When a formula is given, the paper which is recommended for that particular formula should alone be used.

SECTION II.

Positive Printing by Development.

Negative printing processes will be found useful during the dull winter months, and at other times when the light is feeble, or when it is required to produce a large number of impressions from a Negative in a short space of time. The plan of development also enables the operator to obtain Positives of greater stability than those yielded by the direct action of light.

Three processes may be described, the first of which gives Positives of an agreeable colour, but the second, on Iodide of Silver, the greatest permanency under unfavourable conditions.

NEGATIVE PRINTING PROCESSES UPON CHLORIDE OF
SILVER.

Positives may be obtained by exposing paper prepared with Chloride of Silver to the action of light until a faint image is perceptible, and subsequently developing by Gallic Acid; but in this process it is difficult to obtain sufficient *contrast* of light and shade; the impression, if sufficiently exposed and not too much developed, being feeble, with a want of intensity in the dark parts. By associating with the Chloride an organic salt of Silver, such as the Citrate, this difficulty may be overcome, and the shadows be brought out with great depth and distinctness.

The papers are salted with a mixed Chloride and Citrate as in the formula for the Ammonio-Nitrate Process.* They are then rendered sensitive upon a Bath of Nitrate of Silver *containing* either Citric or *Acetic Acid*, which are used in Negative processes to preserve the clearness of the white parts under the influence of the developer.

The Bath of Aceto-Nitrate is prepared as follows :—

Nitrate of Silver . . .	30 grains.
Glacial Acetic Acid . .	30 minims.
Water	1 fluid ounce.

Float the papers (Papier Saxe or Papier Rive) upon the Bath for three minutes, and suspend them to dry in a room from which actinic rays are *perfectly* excluded.

The exposure to light,—which is conducted in the ordinary printing frame, the Negative and sensitive paper being laid in contact in the usual way,—will seldom be longer than three or four minutes, even upon a dull day, It may be regulated by the colour assumed by the projecting margin of the paper ; but it is quite possible to tell by the appearance of the image when it has received a sufficient amount of exposure :—the whole of the picture should be seen,

---

* The formula at p. 246 may be modified with advantage : use double the quantity of Gelatine, and half the amount of Citrate and Chloride.

excepting the *lightest shades,* and it will be found that very few details can be brought out in the development which were altogether invisible before the Gallic Acid was applied.

The developing solution is prepared as follows :—

> Gallic Acid . . . . . 2 grains.
> Water . . . . 1 fluid ounce.

In very cold weather it may be necessary to employ a saturated solution of Gallic Acid, containing about four grains to the ounce; whereas in warm weather the image will develope too quickly, and Acetic Acid must be added (see the remarks at the end of the process, p. 266).

To facilitate the solution of the Gallic Acid, stand the bottle in a warm place near the fire. A lump of Camphor floated in the liquid, or a drop of Oil of Cloves added, will to a great extent prevent it from becoming mouldy by keeping; but if once mould has formed, the bottle must be well cleansed with Nitric Acid, or the decomposition of the fresh Gallic Acid will be hastened.

Pour the solution of Gallic Acid into a flat dish, and immerse the Prints two or three at a time, moving them about, and using a glass rod to remove air-bubbles. The development is rapid, and will be completed in three or four minutes. If the Print developes slowly, becomes *very dark in colour* by continuing the action of the Gallic Acid, but shows no half-tones, it has not been exposed sufficiently long to the light. An *over*-exposed proof, on the other hand, developes with unusual rapidity, and it is necessary to remove it speedily from the Bath in order to preserve the clearness of the white parts; when taken out to the light, it appears pale and red, with no depth of shadow.

The extent to which the development should be carried depends upon the kind of Print desired. By pushing the action of the Gallic Acid, a dark picture not much altered by the fixing Bath will be produced. But a better result as regards colour and gradation of tone will be obtained by removing the Print from the developing solution whilst in the light red stage, and toning it subsequently by means of Gold; in which case it will correspond both in appearance and properties to a Positive obtained by the direct action of light (see the remarks at page 167).

When it is intended to follow the latter plan, the action of the developer must be stopped at a point when the proof appears lighter than it is to remain; since the Sel d'or Bath adds a little to the intensity, and the image becomes somewhat more vigorous on drying.

Wash the Prints in cold water in order to extract all the Gallic Acid. Then tone with *Sel d'or* in the manner described in the next Section, and fix in the usual way. The whites will with care be kept pure; or with only a faint yellow tinge, which is not objectionable.

Upon comparing the developed Prints with others obtained by the direct action of light upon the same sensitive paper, it is evident that the advantage is *slightly* on the side of the latter; but the difference is so small that it would be overlooked in printing large subjects, for which the Negative Process is more especially adapted. The *colour* of both kinds of Positives is the same, or perhaps a shade darker in the developed proofs, which are usually of a violet-purple tone, but sometimes of a dark chocolate-brown.

A developing process with Serum of Milk.—The use of "whey" as a vehicle for Chloride of Silver has something the same effect as that produced by adding a Citrate. This may be traced to the presence of the Milk Sugar and of a portion of uncoagulated Caseine left in the Serum.

The only difficulty in the process is to coagulate the milk in such a way as to separate the greater part but not the whole of the Caseine. Milk which has become sour, or to which an acid has been added, is not considered so good for the purpose as that which has been treated with rennet; and even when rennet is used it must be of the best quality or its action will be imperfect. The serum must filter clear through blotting-paper; but it should not run very rapidly, or in all probability the whole of the Caseine has been separated, and the fluid contains little besides sugar. The whey which is left after cheese-making, commonly answers the purpose, if clarified by beating it up with the white of an egg and subsequently boiling and filtering. Globules of oil must be separated as far as possible, or they will produce a greasiness of the paper.*

Salt the prepared Serum with Chloride of Sodium or Ammonium; in quantity about eight or ten grains to each fluid ounce, and render sensitive upon the same Bath as that recommended for the Citrate Process.

A NEGATIVE PRINTING PROCESS UPON IODIDE OF SILVER.

Iodide of Silver is more sensitive to the reception of the invisible image than the other compounds of that metal; and hence it is usefully employed in printing *enlarged* Positives from small Negatives, by means of the Camera. The great stability of the proofs upon Iodide of Silver will also be a recommendation of this process when unusual permanency is required.

Take of

> Iodide of Potassium . . 160 grains.
> Water . . . . 20 fluid ounces.

The best paper to use will be either Turner's Calotype, or Whatman's or Hollingworth's Negative; the foreign papers do not succeed with the above formula (p. 258).

Float the paper on the iodizing Bath until it ceases to curl up and lies flat upon the liquid: then pin up to dry in the usual way.

Render sensitive upon a Bath of Aceto-Nitrate of Silver containing 30 grains of Nitrate of Silver with 30 minims of Glacial Acetic Acid to each ounce of water.

When the sheet is quite dry, place it in contact with the

---

* See the Vocabulary, Part III., Art. "Milk," for further particulars.

Negative in a pressure frame, and expose *to a feeble light*. About 30 seconds will be an average time upon a dull winter's day, on which it would be impossible to print at all in the ordinary way. On removing the Negative nothing whatever is seen upon the paper, the image being strictly invisible in this process unless the exposure has been carried too far.

Develope by immersion in a saturated solution of Gallic Acid, prepared in the manner described at page 261. The image appears slowly, and the process may last from 15 minutes to half an hour. If the exposure has been correctly timed, the Gallic Acid appears at length almost to cease acting; but when the proof has been over-exposed, the development goes on uninterruptedly, and the image becomes too dark, partaking more of the character of a Negative than a Positive. The usual rule, that *under*-exposed proofs develope slowly but show no half-tones, and that the *over*-exposed develope with unusual rapidity, is also observed in the process with Iodide of Silver.

After the picture is fully brought out, wash in cold, and subsequently in warm water, to remove the Gallic Acid, which, if allowed to remain, would discolour the Hyposulphite Bath. Then fix the Print in a solution of Hyposulphite of Soda, one part to two of water, continuing the action until the yellow colour of the Iodide disappears. The fixing Bath ought not to produce much change in the tint. If the Positive loses its dark colour on immersion in the Hyposulphite, and becomes pale and red, it has been insufficiently developed. The theory of this part of the process should be understood:—It is particularly the *second stage* of the development of a Photograph (see p. 144) on which the fixing Bath produces no effect; and therefore a considerable change of colour in the Hyposulphite indicates that too little Silver has been deposited, and the remedy will be to push the development, adding a little Aceto-Nitrate to the Gallic Acid if the strength of the Bath be found insufficient to yield dark tones.

. The colour of Positives developed upon Iodide of Silver is not agreeable, and they become blue and inky when toned with gold. By fixing the proof in Hyposulphite of Soda which has been long used and has acquired sulphuretting properties, the tint is much improved ; but the permanency of the Print under unfavourable conditions is lessened by adopting that mode of toning.

A NEGATIVE PRINTING PROCESS UPON BROMIDE OF SILVER.

By substituting the Bromide for the Iodide of Silver in the above process, the proportions and details of manipulation being in other respects the same, a more agreeable colour is obtained.

Paper prepared with Bromide of Silver is less sensitive than the Iodide, but an exposure of one minute (in the printing frame) will usually be sufficient even on a dull day. The image is nearly latent, but sometimes a very faint outline of the darkest shadows can be seen. The proportion of Bromide used is likely to influence this point; the sensitiveness being diminished, but the image showing more of the details before development, when the quantity of the Silver Salt is reduced to a minimum.

Either English or French papers may be used, but in the latter case the Bromide should be dissolved in Serum of Milk (p. 262), or it will be difficult to obtain a good surface picture. The proportion of Bromide may be five grains to the ounce of Serum.

These proofs, even when simply fixed in plain Hyposulphite of Soda, are superior in colour to the Positives printed by the last formula upon Iodide of Silver ; and the permanency is very great if the development be sufficiently pushed. The use of the Serum of Milk gives an advantage in resisting the oxidizing influences to which Positives are liable to be exposed (p. 150).

GENERAL REMARKS ON NEGATIVE PRINTING.

Printing by development should not be attempted until

the manipulation of the ordinary process by direct exposure to light has been acquired.

Perfect cleanliness is essential. The salting or iodizing solution and the Aceto-Nitrate Bath must be filtered clear, as the effect of small suspended particles in producing spots is more seen when the image is brought out by a developer.

It will be necessary to be far more careful in excluding white light than in the ordinary process; and when Iodide of Silver is used, all the precautions required in the case of Collodion Negatives must be taken.

Observe particularly that the dishes are kept clean, or the Gallo-Nitrate of Silver will be rapidly discoloured (read the remarks at page 179).

Stereoscopic Negatives and small portraits are not successfully printed by development; since it is difficult to obtain the most elaborate definition, and there is a slight tendency to yellowness in the white parts. Positives may be developed upon Albumen paper, but the Gallic Acid is apt to discolour the lights.

In printing by development upon Chloride of Silver, the theory of the subject must be particularly studied. When the weather is cold and the light bad, the development of the image proceeds slowly, the Gallic Acid Bath remains clear, and good half-tones are obtained; but under opposite conditions, the developer may become turbid and the shadows be lost by excessive deposit of Silver. This *over-development* will be remedied by printing the Negative in a more feeble light (near to the open window of a room), and by adding Acetic Acid to the developer, about 5 or 10 minims to the ounce, so as to bring out the image more slowly. The intensity of action is thus lessened, and if the picture be not under-exposed, the half-tones will be good.

Observe also when preparing papers with Citrate, that if too much Carbonate of Soda be added in neutralizing the Citric Acid, Carbonate of Silver will be deposited in the paper, the effect of which is to remove by degrees the acidity of the Nitrate Bath, and to produce over-development and excessive sensibility to light.

The colour of the proofs when taken from the Gallic Acid should be *light red;* the gradation of tone not being usually so perfect when the development is carried into the second or black stage.

· It is not recommended to prepare too large a stock of the salted papers, as they will probably be liable to mouldiness and decomposition unless kept perfectly dry.

SECTION III.

The Sel d'or Process for toning Positives.

This process is somewhat more troublesome than the plan of fixing and toning in one solution, but possesses advantages which will presently be enumerated. The description may be divided into the preparation of the toning Bath, and the manipulatory details.

THE PREPARATION OF THE TONING BATH.

Take of

Chloride of Gold	1 grain.
Pure Hyposulphite of Soda .	3 grains.
Hydrochloric Acid	4 minims.
Water, distilled or common .	4 fluid ounces.

Dissolve the Gold and Hyposulphite of Soda each in two ounces of the water; then mix quickly by pouring the former solution into the latter, and add the Hydrochloric Acid. If the Chloride of Gold be neutral, the liquid will have a red tinge, but if *acid*, then the solution may be colourless. The commercial Chloride of Gold, containing usually much free Hydrochloric Acid, will not require any addition of that substance. (See the Vocabulary, Part III.)

In place of making an extemporaneous Hyposulphite of Gold by mixing the Chloride with Hyposulphite of Soda, the Crystallized Sel d'or may be used, adding about half a grain to the ounce of water, acidified as before; but the objection to the employment of this salt is its expense, and also the difficulty of obtaining it in a pure form; some samples containing less than five per cent. of Gold.

It will be found very convenient to keep the two solutions on hand ready for mixing, viz. the Chloride of Gold dissolved in water in the proportion of a grain to the drachm, and the Hyposulphite of Soda, three grains to the drachm. When required for use, measure out a fluid drachm of each, dilute with water to two ounces, and mix.

It is possible that the three-grain solution of Hyposulphite of Soda may by long keeping become decomposed, with precipitation of Sulphur. The effect of this would be to produce a turbidity and deposit of Gold on mixing the ingredients for the Bath, the Chloride of Gold being in excess over the Hyposulphite of Soda (see p. 250).

The Bath of Sel d'or is always most active when recently mixed, but it will keep good for some days if contact with free Nitrate of Silver be avoided. The addition of this substance produces a red deposit in the Bath, containing Gold, and the solution then becomes useless.

DETAILS OF MANIPULATION.

The paper may be prepared by either of the formulæ given in the first Section of this Chapter, according to the tint desired. The pure black tones are obtained most easily with the Ammonio-Nitrate paper, and the purple tints, without gloss, on paper prepared with plain Chloride and Citrate of Soda.

The printing is not carried quite to the usual intensity, as the half-tones are very little dissolved in this process.

On being taken from the frame, the prints are washed thoroughly in common water until it ceases to become milky; that is, until the greater part of the Nitrate of Silver has been removed. The washing must be conducted in a dark place, but it is not necessary to hasten it; the proofs may be thrown into a pan of water covered with a cloth, and allowed to remain until required for tinting.

A trace of free Nitrate of Silver usually escapes the washing; this would cause a yellow deposit on the Print, and also in the toning Bath. It must therefore be removed, either by adding a little *common salt* to the water during the last washings, or by means of a dilute solution of Ammonia.

For plain paper Prints the former plan will be found the least troublesome; but with Albumen proofs* the Ammonia is required, in order to dissolve away a portion of the Albuminate of Silver which has escaped the action of light, before submitting the print to the gold; otherwise the dark tones would nearly disappear in the fixing Bath, the Hyposulphite carrying away the Gold with this superficial layer of silver salt.

To prepare the Ammonia Bath, take of

> Liquor Ammoniæ . . . . 1 drachm.
> Common Water . . . . 1 pint.

The exact quantity is not material; if the liquid smells faintly of Ammonia, it will be sufficient. Place the washed Prints in this Bath, two or three at a time, and allow them to remain until the purple tint gives place to a red tone. The action must be watched, because if the Ammonia Bath be strong, the proof becomes unusually *pale and red*, and when this is the case a little brilliancy is lost in the after-tinting.

As the Print is comparatively insensitive to light when the excess of Nitrate has been washed away, it is not necessary to darken the room; but a *bright light* proceeding from an open door or window should be avoided.

* The amateur is recommended not to use Albuminized paper in this process until he has become accustomed to the manipulations; the plain paper prints being toned with more ease and certainty.

After using the salt or the Ammonia, soak the Prints again for a minute or so in common water. Then place them in the toning Bath of Gold and acid; do not put in too many at once, and move them about occasionally, to prevent spots of imperfect action at the point where the sheets touch each other.

The foreign papers, plain salted, colour rapidly in two or three minutes. English papers require five to ten minutes; Albuminized, ten minutes to a quarter of an hour. The tendency of the Gold Bath is to give a blue tone to the image; hence proofs which are light red after using the salt or Ammonia, become, first red-purple, and then violet-purple in the Sel d'or. Albumen Prints assume some shade of brown, or of purple if not too strongly Albuminized. Ammonio-Nitrate papers highly salted, and prepared without Citrate, become first dark purple, and then blue and inky; the Citrate is intended to obviate this inky tint.

When the darkest tones are reached, the Bath produces no further effect, but eventually (more especially if the solution be not shielded from light [?]) there is a little decomposition, producing a cream-coloured deposit upon the lights.

The toning being completed, the Prints are again washed for an instant in water, to remove the excess of gold solution. This washing must not be continued longer than two or three minutes, or there will be danger of yellowness of the whites; this however ought not to happen with proper precautions.

Lastly, the proofs are fixed in a solution of Hyposulphite of Soda, one part to four of water; which may be used many times successively. This Bath alters the tone very little if the deposit of Gold be well fixed on the Print; but the writer has often observed in the case of Albumen paper and paper prepared with Citrate (Formula II.) that if removed too quickly from the Sel d'or, the purple tones change by immersion in the Hyposulphite to a chocolate-brown. Ammonio-Nitrate Prints are less liable to alter in this way.

In order that the fixing may be properly performed, the time of immersion should not be less than ten minutes with a porous paper, plain salted; or fifteen minutes in the case of an English or albuminized paper.

Ammonia may be used for fixing plain paper Prints; about one part of the Liquor Ammoniæ, to four of water. Ten minutes' immersion will usually be sufficient, and the tone is very little affected. This process is a good one, but the pungent smell of the Ammonia is an objection, and the Bath discolours by use. Some care too is required in order to ensure a proper fixing of the prints (see the remarks at page 131).

For directions to wash and mount the proofs, see page 255.

It will sometimes happen in the Sel d'or process, from the toning Bath having but little solvent action on the light shades, that the Prints, after being washed and dried, appear too dark; this may be remedied by laying them for a few minutes in *a very dilute solution* of Chloride of Gold (five or six drops of 'the yellow solution of the Chloride to a few ounces of water) and washing for an additional quarter of an hour. Or an over-printed Positive may be saved by toning it with Chloride of Gold instead of Sel d'or. In that case, after proper removal of the free Nitrate of Silver, a few drops of a lemon-yellow solution of Chloride of Gold (with a fragment of Carbonate of Soda added to remove acidity, p. 132), should be poured over the Print, which is to be subsequently fixed in the usual way.

Advantages of toning by Sel d'or.—This process will be found especially useful by those who print large Positives, The solutions may be mixed in a few minutes, and, being very dilute, are economical. It is not even necessary to employ a *Bath* for toning, but if the Sel d'or solution be prepared of about twice or three times the strength given in the formula, it will be sufficient to pour a few drachms upon the surface of the print. As the Gold solution is always used soon after mixing, a uniform and permanent tint can be obtained ; whereas the single fixing and toning Bath of Gold and Hyposulphite loses much of its efficacy by keeping, and *over-printing* of the proof is required in proportion as the Bath becomes older.

SECTION IV.

On a mode of Printing enlarged and reduced Positives, Transparencies, etc., from Collodion Negatives.

To explain the manner in which a Photograph' may be enlarged or reduced in the process of printing, it will be necessary to refer to the remarks made at page 52, on the *conjugate foci* of lenses.

If a Collodion Negative be placed at a certain distance in front of a Camera, and (by using a tube of black cloth) the light be admitted into the dark chamber only through the Negative, a reduced image will be formed upon the ground glass ; but if the Negative be advanced nearer, the image will increase in size, until it becomes first equal to, and then larger than, the original Negative ; the focus becoming more and more distant from the lens, or *receding*, as the Negative is brought nearer.

Again, if a Negative portrait be placed in the Camera slide, and the instrument being carried into a dark room, a hole be cut in the window-shutter so as to admit light through the Negative, the luminous rays, after refraction by the lens, will form an image of the exact size of life upon a white screen placed in the position originally occupied by the sitter. These two planes, in fact, that of the object and of the image, are strictly *conjugate foci*, and, as regards the result, it is immaterial from which of the two, anterior or posterior, the rays of light proceed.

Therefore in order to obtain a reduced or enlarged copy of a Negative, it is necessary only to form an image of the size required, and to project the image upon a sensitive surface either of Collodion or paper.

A good arrangement for this purpose may be made by taking an ordinary Portrait Camera, and prolonging it in front by a deal box blackened inside and with a double body, to admit of being lengthened out as required; or, more simply, by adding a framework of wood covered in with black cloth. A groove in front carries the Negative, or receives the slide containing the sensitive layer, as the case may be.

In *reducing* Photographs, the Negative is placed in front of the lens, in the position ordinarily occupied by the object; but in making an enlarged copy, it must be fixed *behind* the lens, or, which is equivalent, the lens must be turned round, so that the rays of light transmitted by the Negative enter the back glass of the combination, and pass out at the front. This point should be attended to in order to avoid indistinctness of image from spherical aberration.

A Portrait combination of lenses of $2\frac{1}{2}$ or $3\frac{1}{4}$ inches diameter is the best form to use, and the actinic and luminous foci should accurately correspond, as any difference between them would be increased by enlarging. A stop of an inch or an inch and a half aperture placed *between* the lenses obviates to some extent the loss of sharp outline usually following enlargement of the image.

The light may be admitted through the Negative by pointing the Camera towards the sky; or direct sunlight may be used, thrown upon the Negative by a plane reflector. A common swing looking-glass, if clear and free from specks, does very well; it should be so placed that the centre on which it turns is on a level with the axis of the lens.

The best Negatives for printing enlarged Positives are those which are distinct and clear; and it is important to use a *small* Negative, which strains the lens less and gives a better result than one of larger size. In printing by a $2\frac{1}{4}$ lens for instance, prepare the Negative upon a plate about two inches square, and afterwards enlarge it four diameters.

Paper containing Chloride of Silver is not sufficiently

T

sensitive to receive the image, and the Print should be
formed upon Collodion, or on iodized paper developed by
Gallic Acid (see p. 263).

The exposure required will vary not only with the intensity of the light and the sensibility of the surface used,
but also *with the degree of reduction or enlargement of the
image*.

In printing upon Collodion the resulting picture is
Positive by transmitted light; it should be backed up
with white varnish, and then becomes Positive by reflected
light.   The tone of the blacks is improved by treating the
plate first with Bichloride of Mercury, and then with Ammonia, in the manner described at pages 113 and 207).

Mr. Wenham, who has written a paper on the mode of
obtaining Positives of the life size, operates in the following way:—he places the Camera, with the slide containing
the Negative, in a dark room, and reflects the sunlight in
through a hole in the shutter, so as to pass first through
the Negative and then through the lens; the image is received upon iodized paper, and developed by Gallic Acid,
in the mode described in the second Section of this Chapter (p. 263).

On printing Collodion transparencies for the Stereoscope.
—This may be done by using the Camera to form an image
of the Negative in the mode described in the last page;
but more simply by the following process:—Coat the glass,
upon which the Print is to be formed, with Collodio-Iodide
of Silver in the usual way; then lay it upon a piece of black
cloth, Collodion side uppermost, and place two strips of
paper of about the thickness of cardboard and one-fourth
of an inch broad, along the two opposite edges, to prevent
the Negative being soiled by contact with the film.   Both
glasses must be *perfectly flat*, and even then it may happen
that the Negative is unavoidably wetted; if so, wash it
immediately with water, and if it be properly varnished,
no harm will result.

A little ingenuity will suggest a simple framework of

wood, on which the Negative and sensitive plate are retained, separated only by the thickness of a sheet of paper; and the use of this will be better than holding the combination in the hand.

The printing is conducted by the light of gas, or of a camphine or moderator lamp; diffused daylight would be too powerful.

The employment of a concave reflector, which may be purchased for a few shillings, ensures parallelism of rays, and is a great improvement. The lamp is placed in the focus of the mirror, which may at once be ascertained by moving it backwards and forwards until *an evenly illuminated circle* is thrown upon a white screen held in front. This in fact is one of the disadvantages of printing by a naked flame—that the light falls most powerfully upon the central part, and less so upon the edges, of the Negative.

The picture must be exposed for a longer or shorter time (about ten seconds will be an average) according to its behaviour during development (see p. 224); this process, as well as the fixing, is conducted in the same manner as for Collodion pictures generally.

Some adopt the plan of whitening by Corrosive Sublimate, and again blackening by dilute Ammonia, as an improvement to the colour of the dark shadows (see p. 113).

If this mode of printing upon Collodion be conducted with care, the Negative being separated from the film by the smallest interval only, the loss of distinctness in outline will scarcely be perceived.

Stereoscopic transparencies may also be printed by the dry Collodion process described in Chapter VI., or by the Collodio-Albumen process. Mr. Llewellyn recommends the employment of a solution of Oxymel, so dilute that the plate becomes nearly dry, and may be laid in contact with the Negative without fear of injury (see the footnote at page 302).

CHAPTER V.

CLASSIFICATION OF CAUSES OF FAILURE IN THE COLLODION PROCESS.

Section I.—Imperfections in Collodion Photographs.
Section II.—Imperfections in Paper Positives.

SECTION I.

Imperfections in Negative and Positive Collodion Photographs.

The following may be mentioned:—fogging—spots—markings, etc.

CAUSES OF FOGGING OF COLLODION PLATES.

1. *Over-exposure of the Plate.*—This is likely to happen when using the full aperture of a double combination lens for distant objects brightly illuminated, the Collodion being highly sensitive. Also from the film being very blue and transparent, with too little Iodide of Silver (p. 114).

2. *Diffused Light.*—*a.* In the developing room. This is a frequent cause of fogging, and especially so when the common yellow calico is employed, which is apt to fade. Use a treble thickness, or procure the waterproof material, in which the pores are stopped with gutta-percha. —*b.* In the Camera. The slide may not fit accurately, or

the door does not shut close. Throw a black cloth over the Camera during the exposure of the plate.—*c*. From direct rays of the sun or the light of the sky falling upon the lens. With the full aperture of a double combination Lens, a portion of sky included in the field (as for instance to form the background of a portrait) is apt to cause fogging. The portrait will probably be more brilliant if a funnel-shaped canvas bag, or a curtain with an oblong aperture admitting only the rays proceeding from the sitter, be placed in front of the Camera.

3. *Alkalinity of the Bath.*—This condition, explained at page 88, may be due to one of the following causes:—*a*. The use of Nitrate of Silver which has been too strongly fused (p. 13).—*b*. Constant employment of a Collodion containing free Ammonia or Carbonate of Ammonia (p. 89). —*c*. Addition of Potash, Ammonia, or Carbonate of Soda to the Nitrate Bath, in order to remove free Nitric Acid (p. 89).—*d*. Use of rain-water or hard water for making the Nitrate Bath (rain-water usually contains traces of Ammonia; hard water often abounds with Carbonate of Lime).

In either case the alkalinity may easily be removed by the addition of Acetic Acid, one drop to four ounces of the solution. The proper mode of testing for alkalinity is described at p. 89.

4. *Decomposition of the Nitrate Bath.*—*a*. By constant exposure to light (the injurious effects of this will be mostly seen when Positives are taken).—*b*. By organic matter: this is sometimes present in Nitrate of Silver which has been prepared from the residues of old Baths; or it may be introduced by floating papers for the printing process upon the Bath, or by dissolving the crystals of Nitrate of Silver in *putrid* rain-water, or in *impure* distilled water collected from the condensed water of steam-boilers and contaminated with oily matter.—*c*. Decomposition of the Bath by contact with metallic iron or copper, or with a fixing agent, or a developing agent (p. 90).

5. *Faults of the developing solution.*—*a*. Brown and de-

composed solution of Pyrogallic Acid; this may sometimes be used with impunity, but it tends, as a rule, to facilitate irregular reduction of Silver.—*b.* Impure Acetic Acid having a smell of Garlic and which probably contains Sulphur in organic combination.—*c.* Omission of the Acetic Acid in the developer: this will produce a universal blackness.

6. *Sundry other causes of fogging.*—*a.* Vapour of Ammonia or Hydrosulphate of Ammonia, or the products of the combustion of coal-gas, escaping into the developing room.—*b.* Development of the image by *immersion* in solution of Sulphate of Iron: this is a safe plan when the films are formed in an acid Nitrate Bath; but with pale films formed in a chemically neutral Bath it is better to pour the fluid over the plate, and not to use the same portion twice.—*c.* Redipping the plate in the Bath before development: this is apt to give a foggy picture when using an old Bath, and is not recommended.

Systematic plan of proceeding to detect the cause of the fogging.—If the amateur has had but little experience in the Collodion process, and is using Collodion of moderate sensitiveness and a new Bath, the probability is that the fogging is caused by *over-exposure.* Having obviated this, proceed to test the Bath; *if it is made from pure materials, and does not restore the blue colour of a piece of litmus-paper previously reddened by holding it over the mouth of a glacial Acetic Acid bottle,* it may be considered in working order.

Next prepare a sensitive plate, and after draining it for two or three minutes in a dark place, pour on the developer: wash, fix, and bring out to the light; if any mistiness is perceptible, either the developing room is not sufficiently dark, or the Bath was prepared with a bad sample of Nitrate of Silver, or with impure Alcohol, or impure water.

On the other hand, if the plate remains absolutely clear under these circumstances, *the cause of error may be in the Camera;*—therefore prepare another sensitive film, place

it in the Camera, and proceed exactly as if taking a picture, with the exception of not removing the brass cap of
the lens: allow to remain for two or three minutes, and
then remove and develope as usual.

If no indication of the cause of the fogging is obtained
in either of these ways, there is every reason to suppose
that it is due to diffused Light gaining entrance through
the lens. This cause of error may often be detected by
looking into the Camera from the front, when an irregular
reflection will be seen upon the glass.

SPOTS UPON COLLODION PLATES.

Spots are of two kinds: spots of *opacity*, which appear
black by transmitted light, and *white* by reflected light;
and spots of *transparency*, the reverse of the others, being
white when seen upon Negatives, and black on Positives.

OPAQUE SPOTS are referable to *an excess of development* at the point where the spot is seen; they may be
caused by—

1. *The use of Collodion holding small particles in suspension.*—Each particle becomes a centre of chemical action, and produces a speck, or a speck with a tail to it.
The Collodion should be placed aside to settle for several
hours, after which the upper portion may be poured off.

2. *Turbidity of the Nitrate solution.*—*a.* From flakes of
Iodide of Silver having fallen away into the solution, by
use of an over-iodized Collodion.—*b.* From a deposit formed
by degrees upon the sides of the gutta-percha trough.—
c. From the inside of the trough being *dusty* at the time
of pouring in the solution.

In order to obviate these inconveniences, it is well to
make at least half as much again of the Nitrate solution
as is necessary, and to keep it in a stock-bottle, from which
the upper part may be poured off when required. The
frequent filtration of Silver Baths is unadvisable, since the
paper employed may be contaminated with impurities.

3. *Dust upon the surface of the glass at the time of pouring on the Collodion.*—Perfectly clean glasses, if set aside for a few minutes, acquire small particles of dust; each plate should therefore be gently wiped with a silk handkerchief immediately before being used.

4. *Faults of the Slide.*—Sometimes a small hole exists, which admits a pencil of light, and produces a spot, known by its being always in the same part of the plate; occasionally the door works too tightly, so that small particles of wood, etc., are scraped off, and projected against the plate when it is raised. Or perhaps the operator, after the exposure is finished, shuts down the door with a jerk, and so causes a *splash* in the liquid which has drained down and accumulated in the groove below; this cause, although not a common one, may sometimes occur.

5. *Insoluble particles in the Pyrogallic Acid.*—The solution of Pyrogallic Acid will not usually require filtering, but if specks of Metagallic Acid are present, the developer should be passed through blotting paper before use.

SPOTS OF TRANSPARENCY may generally be traced to some cause *which renders the Iodide of Silver insensible to light at particular points*, so that on the application of the developer no reduction takes place.

1. *Concentration of the Nitrate of Silver on the surface of the film by evaporation.*—When the film becomes too dry after removal from the Bath, the solvent power of the Nitrate increases so much that it eats away the Iodide and produces spots.

2. *Small particles of undissolved Iodide of Potassium in the Collodion.*—These are likely to occur when Anhydrous Ether and Alcohol are employed. They produce transparent specks at every part of the plate. Allow the Collodion to settle, or add a drop of water, which will dissolve the Iodide.

3. *Alcohol or Ether containing too much water.*—This causes a reticulated appearance of the film, which is rotten and full of holes.

4. *Use of glasses improperly cleaned.*—This cause is perhaps the most frequent of all, when the film of Pyroxyline
is very thin and the Bath neutral. After glasses have been
long used it is often difficult to clean them so thoroughly
that the breath lies smoothly; but the use of Potash gives
the best chance.

MARKINGS OF VARIOUS KINDS ON COLLODION PLATES.

1. *A reticulated appearance on the film after developing.*
—When this is *universal*, it often depends upon the employment of Collodion containing water. Or, if not due to
this cause, the plate may have been immersed too quickly
in the Bath, and the soluble Pyroxyline partially precipitated.

2. *Oily spots or lines.*—*a.* From raising the plate out of
the Nitrate Bath before it has been immersed sufficiently
long to have become thoroughly wetted.—*b.* Removal of
the plate from the Bath before the Ether upon the surface
has been washed away.—*c.* Redipping the plate in the Nitrate Bath after exposure to light, and pouring on the developer *immediately;* if a few minutes be not allowed to
drain off the excess of Nitrate, the Pyrogallic Acid will
not amalgamate readily with the surface of the film.—
d. From the Nitrate Bath being covered with an oily scum,
which is carried down by the plate. Draw a slip of blotting-paper gently along the surface of the liquid before
using it.

3. *Straight lines traversing the film horizontally.*—From
a check having been made in immersing the plate in the
Bath.

4. *Curved lines of over-development.*—By employing the
developer too concentrated; or by not pouring it on sufficiently quickly to cover the surface before the action begins; or by using too little Acetic Acid, and omitting the
Alcohol. The addition of Alcohol to the developer will
not be required as a rule when the Bath is newly made;
but when much Ether has accumulated in it, the developer

has a tendency to run into oily lines, unless containing Alcohol.

5. *Stains from too small a quantity of fluid having been employed to develope the image.*—In this case, the whole plate not being thoroughly covered during the development, the action does not always proceed with regularity.

6. *Irregular striæ.*—From fragments of dried Collodion accumulating in the neck of the bottle, and being washed on the film; to avoid this, the finger should be passed gently round the inside of the neck before use.

7. *Markings like those represented in the woodcut.*—They are caused by using an inferior sample of Pyroxyline made from too hot acids, and are most seen when using an old Bath.

8. *Stains on the upper part of the plate, from using a dirty slide.*—To avoid these, place, if necessary, strips of blotting-paper between the supports and the glass.

9. *Wavy marks at the lower parts of the plate.*—*a.* If the Collodion is becoming thick and glutinous from constant use, dilute it with a little Ether containing an eighth part of Alcohol.—*b.* From reversing the direction of the plate after its removal from the Bath, so that the Nitrate of Silver flows back again over the surface and causes a stain on the application of Pyrogallic Acid.—*c.* Impurities on the woodwork of the frame ascending the film by capillary attraction. This is a frequent source of stains.

10. *Marks from the developer not running up to the edge of the film* (p. 212). Remedy this as far as possible by allowing the Collodion to *set* a little more firmly before dipping the plate in the Bath.

IMPERFECTIONS IN COLLODION NEGATIVES.

1. *A want of Intensity.*—*a.* From the development not having been sufficiently pushed (p. 224).—*b.* From the Collodion film being too blue and transparent for Ne-

gatives.—*c*. The Collodion newly made from pure materials (p. 114).—*d*. The plate kept too long between exciting and development (p. 100).—*e*. The Bath newly prepared from commercial crystallized Nitrate of Silver (p. 101).—*f*. The light too feeble, as on very dark wintry days, or in copying interiors, etc.

2. *Inferior half-tones, with great intensity of the high Lights.*—*a*. From the plate being insufficiently exposed. —*b*. The Collodion of inferior quality, either too strongly tinted with Iodine or made from impure materials.—*c*. The Nitrate Bath old and partially decomposed.—*d*. The light reflected too strongly from the object. When the light is unusually bright, a feeble Collodion and a newly mixed Nitrate Bath will be found to give better definition in the high lights than an intense Collodion, which may produce chalky Negatives.

3. *The image pale and misty.*—The plate is over-exposed (if so, the image will probably be a reddish-brown colour by transmitted light), or there is diffused light in the Camera or developing room. The presence of Bromides or Chlorides in the Collodion may occasionally produce the same effect.

4. *The high lights of the image are solarized.*—A change of colour to a light brown or red tint by transmitted light, with a dark shade by reflected light, is favoured by over-exposure of the plate, by organic decomposition of the Collodion, and by Acetate of Silver and other organic bodies in the Bath.

5. *The image dissolves off on applying the Cyanide of Potassium.*—The Collodion is probably over-iodized. The same thing may also happen in the Honey preservative process, when the plates have been long kept and the indurated layer of syrup not properly removed before applying the developer.

6. *The developer does not run up to the edge of the film.*— This is likely to occur when using Collodion nearly anhydrous; and particularly so with a new Bath not contain-

ing much Alcohol. The film will be less repellent, if a longer time be allowed before dipping in the Bath.

7. *The film does not stick to the glass.*—Clean the plates very carefully, and make the Collodion a little thinner if required. Allow a longer time before dipping in the Bath. A very effectual plan is to roughen the surface of the plates, about an eighth of an inch round the edges.

IMPERFECTIONS IN COLLODION POSITIVES.

The principal difficulty in the production of Negatives is to ascertain the right time of exposure to light and the proper point to which to carry the development of the image. A minor amount of fogging, stains, etc., is of less consequence, and will scarcely be noticed in the printing.

With direct Positives however the case is different. The beauty of these pictures depends entirely upon their being clean and brilliant, without fogging, specks, or imperfections of any kind. On the other hand, the exposure and development of Positives is comparatively simple and easily ascertained.

1. *The shadows dark and heavy.*—The plate has not received sufficient exposure in the Camera;—or the film being very transparent and the Silver solution weak, Nitric Acid is present in the Bath, or the Collodion is brown from free Iodine; in the latter case make the Collodion a little thicker, and develope with Sulphate of Iron in preference to Pyrogallic Acid.

2. *The shadows good, but the lights overdone.*—The developing fluid may have been kept on too long; or the object is not properly illuminated (p. 220); or the Collodion is not adapted for Positives.

3. *The high lights pale and flat, the shadows misty.*—The plate is over-exposed. Indistinctness of outline caused by over-exposure is distinguished from that produced by fogging by holding the plate up to the light; in the former case the image shows as a Negative.

If the Collodion is colourless, clearer shadows will pro-

IN THE COLLODION PROCESS.

bably be obtained by dropping in Tincture of Iodine until a yellow colour is produced.

4. *The picture developes slowly; spangles of metallic Silver are formed.*—Too much Nitric Acid is present in proportion to the strength of the Bath, to the amount of Iodide in the film, and to the quantity of Protosalt of Iron in the developer (p. 112).

5. *Circular spots of a black colour after backing up with the varnish.*—These are often caused by lifting the plate too quickly out of the Bath; or by pouring on the developer at one spot, so as to wash away the Nitrate of Silver; or by the use of glasses imperfectly cleaned.

6. *The image becomes metallic on drying.*—If Sulphate of Iron is employed, the solution is too weak, or free Nitric Acid has been added in excess. If Pyrogallic Acid is used to develope, the proportion of Nitric Acid is too great.

7. *A green or blue tint in certain parts of the image.*—This is caused by the deposit of Silver being too scanty, which may happen from over-action of the light, or from the film of Pyroxyline being *very thin;*—if the Collodion is diluted down beyond a certain point, the same quantity of free Nitrate of Silver is not retained upon the surface of the film. Add a few drops of the Bath to the developer before pouring it on the plate.

8. *Vertical lines, and mistiness, on the image.*—If the Bath has been much used, add to it a third part of a simple solution of Nitrate of Silver in water, without any Alcohol or Iodide. Also prepare the developer with addition of Alcohol, to make it flow more readily (p. 211).

SECTION II.

Imperfections in Paper Positives.

1. *The Print marbled and spotty.*—The quality of the paper is often inferior, which causes it to imbibe liquids

unevenly at different points; or the amount of Silver in the Nitrate Bath is insufficient. In this case the spots are often absent at the lower and most depending part of the sheet, where the excess of liquid drains off.

2. *The Print clean on the surface, but spotted when held up to the light.*—In this case the spots are probably due to imperfect fixation (see p. 129).

3. *The Print becomes pale in the Hyposulphite Bath, and has a cold and faded appearance when finished.*—The Chloride of Silver in the paper may have been in excess with regard to the free Nitrate of Silver; which is especially likely if no *bronzing* could be obtained by prolonged action of the light, or if a weak solution of Nitrate of Silver was laid on with a brush, or by a glass rod. Prints formed on paper which has been kept too long after sensitizing present the same appearance, the free Nitrate of Silver having entered into combination with the organic matter.

4. *Yellowness of the light parts of the proof.*—The following causes are likely to produce yellowness :—*acidity* of the fixing and toning Bath (p. 139),—its action continued for too long a time,—the first washings of the proof not performed quickly,—the toning Bath laid aside until it had become decomposed and nearly useless,—the paper kept for several days after sensitizing.

A creamy yellowness is also common in Prints toned by Sel d'or, when the Hydrochloric Acid has been omitted from the formula; the proof exposed to light during the toning and fixing process; or too long a time allowed to elapse between the toning and fixing. It is also more frequently met with on albuminized paper.

5. *Intense bronzing of the deep shadows.*—In this case the Negative is in fault; remedy the evil as far as possible by printing on paper containing but little salt.

6. *The definition of the Print imperfect, the Negative being a good one.*—Much will depend upon the quality of the paper. Towgood's Positive gives good definition. The use of Albumen will be a great advantage. Citrate of Soda (p. 246) will also improve the definition on plain paper.

7. *Markings of a yellow tint in the dark portions of the Positive.*—These are common on Prints toned without Gold; care should be taken not to handle the paper too much, either before or after sensitizing; to wash the prints in a clean vessel; and not to lay them down whilst wet on a wooden table or in contact with anything likely to communicate impurities.

8. *Small specks and spots of different kinds.*—These, when not corresponding to similar marks upon the Negative, are usually due to metallic specks in the paper; or to insoluble particles floating in the bath.

9. *Markings of the brush in Ammonio-Nitrate pictures.*—In this case there is probably an excess of Ammonia, which dissolves the Chloride of Silver. Add a little fresh Nitrate of Silver, or use the Oxide of Silver dissolved in Nitrate of Ammonia (p. 249).

10. *Marbled stains on the surface of the Print.*—Draw a strip of blotting-paper gently over the surface of the Nitrate Bath before sensitizing the paper; and see that the sheet does not touch the bottom of the dish.

11. *Streaks on Albuminized paper.*—Apply the Albumen more rapidly and evenly to the paper. If this does not succeed, add a little Ox Gall (p. 243).

12. *Removal of the Albumen from the paper during sensitizing.*—The Nitrate Bath is probably alkaline (see page 89).

CHAPTER VI.

LANDSCAPE PHOTOGRAPHY ON PRESERVED COLLODION AND COLLODIO-ALBUMEN.

THE Collodion process may be applied with success to landscape Photography; but as the plates become dry and lose their sensitiveness shortly after their removal from the Bath, the operator will require to provide himself with a yellow tent or some portable vehicle in which the operations of sensitizing and developing can be conducted. As it is a point of great importance in the Collodion process that the plate should receive exactly the right amount of exposure in the Camera,—a few seconds more or less sufficing to affect the character of the picture,—many will submit to much trouble and inconvenience in order to have the apparatus complete upon the spot at which the view is taken.

The object of the "Collodion Preservative Processes" is to maintain the sensitiveness of the film for a certain length of time after it has been excited in the Bath. There is some difficulty in doing this, because if the plate be allowed to dry spontaneously, the solution of free Nitrate of Silver upon the surface, becoming concentrated by evaporation, eats away the Iodide of Silver, and produces transparent spots.

Some operators have attempted to use a second plate of glass in such a way as to *enclose* the sensitive film with an intervening stratum of liquid. The difficulty however of separating the glasses again without tearing the film, is considerable.

In the process of Messrs. Spiller and Crookes, the property possessed by certain saline substances of remaining for a long time in a moist condition was turned to account. Such salts are termed " deliquescent," and many of them have so great an attraction for water that they absorb it eagerly from the air: the solution having been formed, the water cannot entirely be driven off except by the application of a considerable heat.

More recently, Honey has been employed by Mr. Shadbolt.* This substance can scarcely be termed deliquescent, but it possesses, like other uncrystallizable sugars, the property of remaining moist and sticky for a long time. Honey is, according to the Author's views, superior to inorganic deliquescent salts as a preservative agent, from its possessing an affinity for Oxides of Silver, and thus acting chemically in communicating organic intensity to the image.—Collodion plates when kept long in a moist and sensitive state often give a pale and blue image, even although the Nitrate of Silver be left upon the film; and neither Nitrate of Magnesia nor Glycerine appears capable of supplying the deficient element, both being nearly or quite indifferent to the Salts of Silver.

THE HONEY AND OXYMEL KEEPING PROCESSES.

When the weather is cool, Collodion plates may be preserved with tolerable certainty for a few hours, by simply applying Honey to them in the state in which they are taken from the Nitrate of Silver Bath.

The best pure Virgin Honey should be obtained by

---

* A claim has lately been advanced by Mr. Maxwell Lyte to be considered as the discoverer of the Honey Process. This gentleman appears to have worked simultaneously with Mr. Shadbolt, and to have anticipated him in publishing; but the object of Mr. Lyte's process was rather to increase the sensibility of the plates than to confer upon them keeping qualities.

U

dripping it immediately from the comb. This point is of importance, since if the sample of honey be of inferior quality, or adulterated, the process may not succeed. The quantity of water to be added will vary with the consistence of the honey, from about an equal bulk to two parts: it should be sufficient to make the preservative solution pass slowly through filtering-paper.

After the plate is removed from the Nitrate Bath, it is to be drained and wiped on the back in the usual way. The Honey is then poured along the edge in such a manner as to form a broad wave.which forces the Nitrate of Silver solution before it and covers the film. Next drain the plate into a measure and pour on a second portion of Honey as before. This second dose may be used again for the first application to the succeeding plate.

Lastly, stand the glass on blotting-paper in a dark place for about a quarter of an hour or twenty minutes, and wipe the lower edge before putting it into the plate box.

The exposure required will probably be about four or five times as long as that for new and sensitive Collodion, or twice as long as the exposure required for old and brown Collodion.

Before applying the developer, immerse the plate in a Bath of rain-water for five minutes, moving it about occasionally to soften the honey. This will probably be sufficient for plates which have not been kept longer than four hours, and beyond that time the process is not considered certain, since the Honey exercises a slow reducing action upon the Nitrate of Silver.

The solution of Pyrogallic Acid may be used of the ordinary strength, with a full dose of Acetic Acid. Only a faint image comes out at first, but on pouring over the plate a fresh portion of the developer with two or three drops of the Nitrate Bath added to each fluid drachm, it may be intensified to any extent.

Fix with Hyposulphite of Soda, and wash in the usual way.

When the process fails, from heat of the weather or other causes, the image will probably be feeble and red by transmitted light, and the shadows defective and misty. This is especially likely to happen when the Nitrate Bath is very old and contains much Acetate of Silver; or when the same portion of Honey is used more than once, and has undergone partial decomposition by the action of the Nitrate of Silver. The use of *pure* Honey, free from mouldiness and fermentation, will, *in cool weather*, almost certainly ensure success.

A modification of the process when the plates are to be kept over four hours.—In this case the whole, or the greater part of the Nitrate of Silver must be removed before applying the preservative agent. Wash the sensitive plate in water in the manner described for the Oxymel process in the next page. Then apply the syrup as before, using it as thick as possible. Honeyed plates, free from Nitrate of Silver, may commonly be kept for five or six days; often much longer. Dr. Mansell, who has employed this process with great success, speaks of *temperature* as a point to be attended to. In hot weather the same length of keeping properties will not be attained.

Use of Oxymel for preserving Collodion plates.—The principal difficulty in the employment of Honey in Photography, is its disposition to ferment, or to become mouldy. Fermentation occurs most readily in a dilute solution, and will be obviated by using the syrup as thick and free from water as possible. Mr. Llewellyn employs "Oxymel," which is a mixture of Honey and Vinegar, as a preservative agent. This substance will keep even in dilute solution for a long time without decomposition; and, being very readily removed from the plates, does not interfere with the development of the image. The preparation of Oxymel is described in the Vocabulary, Part III.; it must be diluted with three or four parts of water, and filtered.

Certain facts to which attention has been lately drawn by Dr. Norris and Mr. Barnes in working with dry Collo-

dion, may be advantageously borne in mind when using
Oxymel; the preservative solution of which is employed in
so dilute a state that the process resembles to a great extent a dry Collodion process. The observations above referred to relate to the quality of the Collodion best adapted
for the purpose, and will be found at page 298, to which
the reader is referred.

The manipulation of the Oxymel process is very simple.
Two flat gutta-percha dishes are provided, the one containing common water and the other diluted and filtered Oxymel. The Collodion plate, on its removal from the Bath,
is placed in the first dish, which is gently tilted up and
down, to wash away the free Nitrate of Silver. In a few
seconds, when the liquid is rendered milky, it is poured
away, and fresh water being introduced, the process is repeated *until the oily lines disappear, and the surface of the
film becomes smooth and glassy.* The plate is then, after
a slight draining, removed to the second tray, and the
Oxymel waved backwards and forwards for about half a
minute, after which the glass is lifted out and placed vertically on blotting-paper, which must be renewed when it
becomes wet and saturated.

The plates may be used any time within a fortnight
from the date of their preparation, and it is not necessary
to develope immediately after the exposure. The sensitiveness will be considerably less than that of fresh Collodion:
from two to five minutes may be allowed with a Stereoscopic view lens having a quarter-inch diaphragm.

Before developing, the film should be gently washed for
a few seconds with common water. Solution of Pyrogallic Acid, of the ordinary strength, but previously mixed
with a portion of the Nitrate Bath solution, one or two
drops to each drachm, may then be poured on in the ordinary way. Use less Nitrate of Silver and more Acetic
Acid in hot weather. When discoloration of the developer
occurs, mix a fresh portion and proceed as before.

PRECAUTIONS TO BE OBSERVED IN KEEPING PROCESSES.

The plates must be roughened at the edges, and also upon the surface, to make the film adhere.

It is advisable to use a tolerably thick Collodion, giving a yellow film ; the pale opalescent films being more easily affected by markings on the glass, and not retaining so much of syrup or Nitrate of Silver upon the surface.

. The room in which the plates are prepared must be carefully guarded from scattered pencils of white light ; the films are exposed to injury from this cause during the whole of the time occupied in applying the preservative syrup ; and hence anything short of absolute chemical darkness will be likely to cause fogging ; especially so when free Nitrate of Silver is left upon the film.

The water used for washing away the free Nitrate of Silver before applying the preserving liquid, need not be distilled.   Common hard water containing Carbonates and Chlorides, and producing *milkiness* with Nitrate of Silver, will often suffice.   The water of the New River and of the River Thames, with which many parts of London are supplied, may certainly be used ; but in the case of a very *hard* water, containing much Sulphate of Lime, it might perhaps be advisable to substitute clean rain-water, free from brown organic discoloration.

The preservative Oxymel must be carefully filtered, and kept *covered*, in order to protect it from dust.   It will also be necessary occasionally, before using it, to run it through a piece of white cambric, to stop back suspended particles, which, if allowed to remain, would be a source of spots.   If it becomes mouldy, or discoloured by Silver, or ferments and evolves gas, throw it away.

After the syrup is applied and the plates are drained, stow them in a grooved box perfectly protected from light ; or place them in slides, which must be kept scrupulously clean, since any trace of impurity would be likely to produce a stain when the plate was left a long time in the slide. If the preserved plates are kept in a cupboard or box, see that no volatile matter, such as Ammonia, coal-gas, etc., can find entrance.

In changing the plates after the exposure in the Camera, use a large bag made of *several thicknesses* of black calico, with a square of yellow calico let in at the top ; an elastic band securing it round the waist.

THE COLLODIO-ALBUMEN PROCESS.

This process, the theory of which has been briefly explained at page 181, is more sensitive than the one last described, and has the additional advantage of giving *dry* plates, which do not attract dust, and are less liable to injury. The details of manipulation are complex, but this inconvenience is not so much felt when preparing a large number of plates.

Cleaning the Glasses.—Success will greatly depend upon the mode in which this part of the process is performed. The layer of Albumen which is applied to the Collodion film tends to swell and to raise the latter in blisters ; the most effectual mode of obviating which will be to clean the glass so that the film adheres with unusual tenacity.

The Liquor Potassæ of the Druggists, diluted with three or four parts of water, and rubbed on the glass by a roll of flannel (page 214), is very effectual. A mixture of Tri-poli-water and Nitric Acid may however, if desired, be substituted :—

> Tripoli . . . . . . 1 drachm.
> Nitric Acid . . . . . 30 minims.
> Water . . . . . . 1 ounce.

Lay the glass flat on a cloth, and rub the surface carefully with a tuft of cotton-wool dipped in the Tripoli ; then, before the cream dries, wipe it off with a second tuft, and polish with a third. Lastly, breathe upon the glass, and having ascertained that it is chemically clean, apply the Collodion.

Coating with Collodion.—Choose a rather thin Collodion which adheres tightly to the glass. A preparation which has been kept a long time after iodizing will usually answer the purpose very well, and, as a rule, a non-contractile, structureless Collodion is better than one which is glutinous and wavy. The degree of sensibility of the Collodion is not thought to have much influence upon the result.

Coating the Plate.—Apply the Collodion in the usual manner, and allow it full time to set perfectly, before dipping in the Bath, in order to favour its adherence to the glass. With Collodion prepared from anhydrous spirits, about half a minute may be given in cool weather.

The Nitrate Bath.—Take of

Fused Nitrate of Silver . . .	40 grains.
Glacial Acetic Acid	30 minims.
Alcohol	20 minims.
Water	1 fluid ounce.

Saturate with Iodide of Silver as described at page 204, and filter. An immersion of one minute will be sufficient; after which, give the plate an up-and-down movement, and wash it in plain water, in the manner advised for the Oxymel preservative process, at page 292. Then stand it on blotting-paper, to drain for a minute or two, wipe the back of the glass, and pour on the Albumen.

This Bath may become discoloured after a time; continue to use it until it is of a dark sherry-colour, and then treat it with "Kaolin," in the manner and with the precautions advised at pages 91 and 245.

The Iodized Albumen.—Procure eggs, fresh laid, or not more than two or three days old. Separate the whites in the same way as for Albuminized paper (p. 241), and mix by the following Formula:—

Albumen	9 fluid ounces.
Water	3 fluid ounces.

Liquor Ammoniæ . . . . . 2 fluid drachms.
Iodide of Potassium . . . . . 48 grains.
Bromide of Potassium . . . . . 12 grains.

The Iodide and Bromide should be free from Carbonate of Potash, which is said to cause pin-holes in the Negatives. To ensure the absence of this salt, dissolve the total quantity of both Iodide and Bromide in the three ounces of water advised in the formula; then, previously to adding the Ammonia and Albumen, introduce *an excessively minute particle of Iodine*, enough barely to colour the liquid. The Iodine decomposes the Carbonate of Potash, but it must not be used in excess, since free Iodine possesses the property of coagulating Albumen. Iodide of Cadmium also coagulates Albumen, so that the Iodides of Potassium and Ammonium are the best.

Having mixed the ingredients in the order above given, introduce them into a bottle, and shake it violently until they have thoroughly amalgamated. Then transfer to a tall narrow jar; allow to settle for twenty-four hours, and draw off the upper clear portion for use. Particulars of this part of the process have already been given under the head of Albuminized Paper, to which the reader is referred (p. 241).

The ammoniacal solution of Albumen may be kept for some time in a stoppered bottle without much decomposition. If mucous threads form in it, filter through fine linen.

Mode of applying the Albumen.—Cover the moist film with the Albumen in the same way as advised for Collodion (p. 216), pouring on at once a sufficient quantity to cause it to spread in an even and undivided sheet; otherwise a veined appearance may be produced, which will show in the development. Return the excess of Albumen into the bottle, and pour it once again upon the plate: the film will remain clear and transparent, if the whole of the Nitrate of Silver has been properly washed away from the

Collodion. Lastly, stand the plate nearly vertically on blotting-paper to dry. This will occupy five or six hours; but the process may be hastened by artificial heat.

After the Albumen solution has been used to coat a number of plates successively, it becomes diluted with water; the result of which is, that unequal intensity of image is produced at the upper and lower edge of the film.

The iodized Albumen plates are at this stage of the process nearly or quite insensitive to light, and may be preserved unchanged for many weeks.

Sensitizing the Albumen film.—When the plate has become thoroughly dry, it is again introduced into the Bath of Aceto-Nitrate of Silver, and allowed to remain for one minute: then washed with water in the same manner as before, but with even greater care, in order to obviate clouding in the development. If blisters should form on drying, it will be found useful to hasten the process by holding the plates to the fire—or a hot iron may be placed in the centre of a covered box and the glasses reared up round the sides. They will thus dry quickly, and there will not be time for the Albumen to swell much by imbibition.

Exposure in the Camera.—This may be performed at any period within a few weeks from the date of preparation of the plates. For a landscape view with a small Stereoscopic single lens, allow about three minutes in the winter, or one minute and a half in the summer.

Development of the image.—This can be deferred as long as fourteen days after the exposure, with successful results. Pour water over the plate until the film is thoroughly wetted; then cover it with a solution of Pyrogallic Acid containing one grain of the acid to the ounce of water, and twenty minims of Glacial Acetic Acid. Two drops of a neutral solution of Nitrate of Silver made with forty grains of Nitrate to the ounce of water must be previously added to each fluid drachm of the Pyrogallic. The development, in the case of a landscape view taken with sun-light, commences almost immediately, and may be com-

pleted in about ten minutes, but the time occupied in developing will vary greatly with the length of exposure, the
quantity of Nitrate of Silver, and the nature of the subject
copied—a badly lighted interior, for instance, often taking
an hour or longer to appear in all its details.   If the developer should discolour before the proper intensity has
been obtained, pour it off and mix a fresh quantity.

Fixing the image.—Hyposulphite of Soda (one ounce to
four of water) will be found preferable to the Cyanide of
Potassium, as the latter has a solvent effect upon the
Albumen.   An unusually long time will be required, as
the fixing agent must penetrate the Albumen, to reach the
Collodion beneath.

Careful washing in water for five or ten minutes removes the excess of Hyposulphite, and the plate may then
be varnished in the usual way.

THE DRY COLLODION PROCESS.

The earlier attempts to employ sensitive Collodion plates
in a desiccated condition were unsuccessful.   The film of
Pyroxyline shrinks on drying, and becomes almost impervious to moisture : hence, the developing solution not penetrating properly, density cannot easily be obtained.   We
are indebted to Dr. Hill Norris, of Birmingham, for establishing the theory of the subject upon a more correct
basis.   He has pointed out the importance of distinguishing
two different conditions of the Collodion surface,* viz. the
contractile, common in newly-mixed Collodion,—and the
short or *powdery,* in Collodion which has been iodized with
the alkaline Iodides, and kept until much Iodine has been
set free.   The latter is the most suitable condition for the
dry process ; and the practical mode of distinguishing between them is by sensitizing a plate and passing the finger
across it ; if it can be easily pushed away in a firm and
connected skin, it will be unfit for the purpose required.
In order still further to preserve the film in a condition

* See these states of the film more fully described at page 83.

permeable by the developer, it is recommended to coat it whilst moist with a solution of Gelatine.

The dry Collodion process, although less sensitive, is more simple than that on Collodio-Albumen, and possesses many of its advantages; but it is less universally applicable, since it depends entirely for success upon the peculiar state of the Collodion, resembling in this respect the Oxymel process already described.

Mode of preparing the plates.—The glasses are coated with the Collodion in the usual way. *Blistering during development* being liable to happen in this process as in the last, every care must be taken to make the films adhere with the greatest possible tenacity, both by cleaning the glasses with extra care (see p. 294), and also by allowing the Collodion to *set* firmly before dipping in the Bath. The plate may be held from twenty to thirty seconds previous to immersion, or even longer, provided the film, when lifted out of the Bath, appears of uniform thickness throughout (see page 218).

The sensitizing having been completed, wash the plates with plain water, exactly in the same way as for Oxymel (p. 292). If Nitrate of Silver be left, clouding will take place in the process of development. After washing, drain for a few seconds, and immerse in the solution of Gelatine.

To prepare this Bath, take of

Nelson's patent Gelatine . . .	128 grains.
Distilled water	14 ounces.
Alcohol	2½ ounces.

Put the Gelatine in the cold water, and allow it a quarter of an hour to soften and swell; it will then readily dissolve on applying a gentle heat. This may be done in a glazed saucepan or a pipkin of earthenware, taking care not to scorch the bottom part by too strong a heat. Next clarify the solution by adding to it, *whilst barely warm*, a teaspoonful of white of egg (previously beaten up with a silver fork), and afterwards heating nearly to the boiling point. The Alcohol must now be added, to facilitate the coagulation of the Albumen. When this takes place and the liquid becomes clear, filter through a clean piece of cambric folded three or four times. If a hot water filtering apparatus can be obtained, the solution may be made to pass through *paper ;* but as it tends to gelatinize on cooling, the ordinary mode of filtration commonly fails. The quantity of Alcohol in the above formula is greater than is usually recommended, allowance having been made for a partial evaporation of the spirit.

The filtered liquid may be poured into a flat porcelain dish, or a vertical trough, but in either case it will be necessary to stand the vessel in warm water, in order to prevent gelatinization.

The Collodion plate, thoroughly washed, is to be immersed in this solution and moved up and down for two or three minutes. It is then removed, drained on blotting-paper, and dried. The use of artificial heat in drying will be found a great advantage; it prevents the gelatine from settling unequally upon the plate. Those who possess an apparatus made purposely for drying plates by hot air, will experience no difficulty, but an ordinary deal trunk may be made to answer, with a little management. Cover the bottom of the box with blotting-paper, and having heated one or two "flat irons," place them in the centre: then range the glasses side by side, with the coated surface looking inwards; in a quarter of an hour, or from that to twenty minutes, the desiccation will be complete. If the Collodion plates are prepared in a room containing a fire, they may be reared up side by side at a distance of two or three feet, and in that way may be safely dried without fear of injury, provided white light be excluded.

When dry they can be stowed away in a box; all the precautions given at page 293 being observed. The sensibility remains good for many days, possibly for weeks or months in cold weather.

Exposure in the Camera.—Allow from four to eight times

the exposure of the most sensitive moist Collodion. On a clear summer's day, a sun-lit view may require one minute or a minute and a half, with a short focus Stereoscopic lens, having a diaphragm of a quarter of an inch diameter. The average time however with the same lens would be about twice as much, viz. three minutes.

Development of the Image.—Make a saturated solution of Gallic Acid in water by the directions given at page 261. Then dissolve forty grains of pure Nitrate of Silver in one ounce of distilled water. Pour into a flat porcelain dish a sufficient quantity of the Gallic Acid solution to flood the plate readily. Then measure it, and to each fluid ounce add *ten minims* of the solution of Silver, or five minims in hot weather. It is important that no discoloration should occur on mixing these liquids together, to obviate which, observe the following precautions :—Clean the porcelain vessel very carefully with Nitric Acid or Cyanide before use. Employ a pure solution of Nitrate of Silver ; and mix it with the Gallic Acid, in preference to adding the Gallic Acid to the Silver solution (read the remarks at p. 179).

The picture may be expected to appear in five or ten minutes, and in one hour, or from that to four hours (p. 298), the development will be complete. It will not be necessary to keep the plates in motion, but simply to lay them side by side in the solution of the Gallic Acid. If in spite of all precautions the developer begins to blacken before the intensity has reached the proper point, it must be poured off and a fresh mixture prepared. This however will not often happen.

Lastly, when a full amount of opacity has been obtained, wash the plate with water, and fix it in a solution of Hyposulphite of Soda, or dilute solution of Cyanide of Potassium.

Failures in the process.—Stains in the development may arise from using dirty dishes, or glasses which have been left in Gallo-Nitrate of Silver and improperly cleaned. It must be borne in mind that these impurities are not visible to the eye, although they produce the effect of discolouring the developer. A thorough cleansing with strong Nitric Acid or Potash will prove a remedy.

Blisters, unless of large size, may often be disregarded, as they disappear on drying. General cloudiness may depend upon the film having been imperfectly washed. Irregular reduction at certain parts may be due to the Gelatine setting before the plate has become dry, or to stains produced by the finger applied to the upper edge of the plate.*

* Since the above was written, Mr. Maxwell Lyte has communicated to the 'Photographic Journal' (vol. iii.) a dry process in which a *modified* Gelatine is used. The change is produced by boiling a solution of gelatine with dilute Sulphuric Acid, which is afterwards neutralized and removed by means of chalk. The result is to destroy the gelatinizing property of the animal substance; the solution retains its fluidity on cooling, and the necessity of employing artificial heat in drying the plates is avoided.

PART III.

OUTLINES OF GENERAL CHEMISTRY.

OUTLINES OF GENERAL CHEMISTRY.

CHAPTER I.

THE CHEMICAL ELEMENTS AND THEIR COMBINATIONS.

THE limits of the present Work allow only of a simple
sketch of the subjects which it is proposed to treat in
this Chapter. Our attention therefore must be confined
to an explanation of certain points which are alluded to
in the First Part of the Work, and without a proper understanding of which it will be impossible for the reader to
make progress.

The following division may be adopted:—The more
important Elementary Bodies, with their symbols and
atomic weights; the Compounds formed by their union;
the class of Salts; illustrations of the nature of Chemical
Affinity; Chemical Nomenclature; Symbolic Notation; the
laws of Combination; the Atomic Theory; the Chemistry
of Organic Bodies.

THE CHEMICAL ELEMENTS, WITH THEIR SYMBOLS AND ATOMIC WEIGHTS.

The class of elementary bodies embraces all those substances which cannot, in the present state of our knowledge, be resolved into simpler forms of matter.

The chemical elements are divided into "metallic" and

x

"non-metallic," according to the possession of certain general characters.

The following are some of the principal non-metallic elements, with the symbols employed to designate them, and their atomic weights :*—

		Symbol.	Atomic Wt.
Gases.	Oxygen	O	8
	Hydrogen	H	1
	Nitrogen	N	14
	Chlorine	Cl	36
Solids.	Iodine	I	126
	Carbon	C	6
	Sulphur	S	16
	Phosphorus	P	32
Liquid.	Bromine	Br	78
Unknown.	Fluorine	F	19

The metallic elements are more numerous. The following list includes only those which are commonly known:—

		Symbol.	Atomic Wt.
Metals of the Alkalies.	Potassium	K	40
	Sodium	Na	24
Metals of the Alkaline Earths.	Barium	Ba	69
	Calcium	Ca	20
	Magnesium	Mg	12
Metals Proper.	Iron	Fe	28
	Zinc	Zn	32
	Cadmium	Cd	56
	Copper	Cu	32
	Lead	Pb	104
	Tin	Sn	59
	Arsenic	As	•75
	Antimony	Sb	129
Noble Metals.	Mercury	Hg	202
	Silver	Ag	108
	Gold	Au	197
	Platinum	Pt	99

* The atomic weights, with the exception of that of Gold, are taken from the last edition of Brande's 'Manual of Chemistry.'

ON THE BINARY COMPOUNDS OF THE ELEMENTS.

Many of the elementary bodies exhibit a strong tendency to combine with each other, and to form *compounds*, which differ in properties from either of their constituent elements. This attraction, which is termed "Chemical Affinity," is exerted principally between bodies which are opposed to each other in their general characters. Thus, taking for example the elements Chlorine and Iodine—they are analogous in their reactions, and therefore there is but little attraction between them, whereas either of the two combines eagerly with Silver, which is an element of a different class. So, again, Sulphur unites with the metals, but two metallic elements are comparatively indifferent to each other.

Oxygen is by far the most important in the list of chemical elements. It combines with all the others, with the single exception, perhaps, of Fluorine. The attraction, or chemical affinity, however, which is exerted, varies much, in different cases. The metals, as a class, are easily oxidized; whilst many of the non-metallic elements, such as Chlorine, Iodine, Bromine, etc., exhibit but little affinity for Oxygen. *Nitrogen* is also a peculiarly negative element, showing little or no tendency to unite with the others.

Classification of binary compounds containing Oxygen.—When one simple element unites with another, the product is termed a "binary" compound.

There are three distinct classes of binary compounds of Oxygen:—Neutral Oxides, basic Oxides, and acid Oxides.

Neutral and basic Oxides. — Take as examples — the Oxide of Hydrogen, or *Water*, a neutral Oxide; the Oxide of Potassium, or *Potash*, a basic Oxide.

Water is termed a *neutral* oxide, because its affinities are low, and it is comparatively indifferent to other bodies. Potash and Oxide of Silver are examples of *basic* oxides; but there is a great difference between the two in chemical energy, the former belonging to a superior class of bases, viz. the alkaline.

By studying the properties of an alkali (such as Potash or Soda) which are familiar to all, we gain a correct notion of the whole class of basic oxides. An alkali is a substance readily soluble in water, and yielding a solution which has a slimy feel from its solvent action upon the skin. It immediately restores the blue colour of reddened litmus, and changes the blue infusion of cabbage to green. Lastly, it is neutralized and loses all its characteristic properties upon the addition of an acid.

The *weaker bases* are, as a rule, sparingly or not at all soluble in water, neither have they the same caustic and solvent action upon the skin; but they restore the colour of reddened litmus, and neutralize acids in the same manner as the more powerful bases, or alkalies.

The ACID *Oxides.*—This class, taking the stronger acids as the type, may be described as follows:—very soluble in water, the solution possessing an intensely sour taste, and a *corroding* rather than a solvent action upon the skin; changes the blue colour of litmus and other vegetable substances to red, and neutralizes the alkalies and basic oxides generally.

Observe however that these properties are possessed in very various degrees by different acids. Prussic Acid and Carbonic Acid, for instance, are not sour to the taste, and being feeble in their reactions, redden litmus scarcely or not at all. All acids however, without any exception, tend to combine with bases and to neutralize themselves; so that this may be said to be the most characteristic property of the class.

Chemical composition of Acid and Basic Oxides contrasted.—It is a law commonly observed, although with many exceptions, that bases are formed by the union of Oxygen with *metals ;* and acids, by Oxygen uniting with *non-metallic elements.* Thus, Sulphuric Acid is a compound of Sulphur and Oxygen; Nitric Acid, of Nitrogen and Oxygen. But the alkali, Potash, is an oxide of the *metal* Potassium; and the oxides of Iron, Silver, Zinc, etc. are bases, and not acids.

Again, the composition of acids and bases is different in another respect; the former invariably contain more Oxygen in proportion to the other element than the latter. Taking the same examples as before, the two classes may be represented thus :—

Acids { Oil of Vitriol, Sulphur 1 atom, Oxygen 3 atoms.
Aqua-fortis, Nitrogen 1 „ Oxygen 5 „

Bases { Oxide of Silver, Silver 1 atom, Oxygen 1 atom.
Oxide of Iron, Iron 1 „ Oxygen 1 „

The class of Hydrogen Acids.—Oxygen is so essentially the element which forms the acidifying principle of acids, that its very name is derived from that fact (οξυς, acid, and γενναω, to generate). Still there are exceptions to this rule, and in some acids *Hydrogen* appears to play the same part; the *Hydracids*, as they are termed, are formed principally by Hydrogen uniting with elements like Chlorine, Bromine, Iodine, Fluorine, etc. Thus, Muriatic or Hydrochloric Acid contains Chlorine and Hydrogen; Hydriodic Acid contains Iodine and Hydrogen.

Observe, however, that the position held by the Hydrogen in these compounds, is different from that of the Oxygen in the " Oxyacids," as regards the number of atoms usually present; thus—

Aqua-fortis = Nitrogen 1 atom, Oxygen 5 atoms,
Muriatic Acid = Chlorine 1 „ Hydrogen 1 atom;

so that the composition of the Hydracids is analogous to the *basic* oxides, in containing a single atom of each constituent.

THE TERNARY COMPOUNDS OF THE ELEMENTS.

As the various elementary substances unite with each other to form Binary Compounds, so these binary compounds again unite and form *Ternary* Compounds.

Compound bodies however do not, as a rule, unite with simple elements. In illustration, take the action of Nitric Acid upon Silver, described at page 12. No effect is produced upon the metal until *Oxygen* is imparted; then the Oxide of Silver so formed dissolves in the Nitric Acid. In other words, it is necessary that a binary compound should be first formed, before the solution can take place. The mutual attraction or chemical affinity exhibited by compound bodies is, as in the case of elements, most strongly marked when the two substances are opposed to each other in their general properties.

Thus, *acids* do not unite with other acids, but they combine instantly with *alkalies;* the two mutually neutralizing each other and forming "a salt."

Salts therefore are ternary compounds produced by the union of acids and bases; common Salt, formed by neutralizing Muriatic Acid with Soda, being taken as the type of the whole class.

General characters of the Salts.—An aqueous solution of Chloride of Sodium, or common Salt, possesses those characters which are usually termed saline; it is neither sour nor corrosive, but, on the other hand, has a cooling agreeable taste. It produces no effect upon litmus and other vegetable colours, and is wanting in those energetic reactions which are characteristic of both acids and alkalies; hence, although formed by the union of two binary compounds, it differs essentially in properties from both.

All salts however do not correspond to this description of the properties of Chloride of Sodium. The Carbonate of Potash, for instance, is an acrid and alkaline salt, and the Nitrate of Iron reddens litmus-paper. A perfectly neutral salt is formed when a strong acid unites with an energetic base; but if, of the two constituents, one is more powerful than the other, the properties of that one are often seen in the resulting salt. Thus the Carbonate of Potash is *alkaline* to test-paper, because the Carbonic

Acid is feeble in its reactions; but if *Nitric Acid* and *Potash* are brought together, then a Nitrate of Potash is produced, which is *neutral* in every sense of the term.

The Chloride of Sodium and salts of a similar kind are freely soluble in water, but all salts are not so. Some dissolve only sparingly, and others not at all. The Chloride and Iodide of Silver are examples of the latter class; they are not bitter and caustic like the Nitrate of Silver, but are perfectly tasteless from being insoluble in the fluids of the mouth.

It is seen therefore from these examples, and many others which might be adduced, that the popular notion of a saline body is far from being correct, and that, in the language of strict definition, any substance is a salt which is produced by the union of an acid with an alkali, independent of the properties it may possess.

Thus, *Cyanide of Potassium* is a true salt, although highly poisonous; Nitrate of Silver is a salt; the green Sulphate of Iron is a salt; so also is Chalk or Carbonate of Lime, which has neither taste, colour, nor smell.

On the "Hydracid" class of Salts.—The distinction between Oxyacids and Hydracids has already been pointed out (p. 309), the latter having been shown to consist of Hydrogen united with elements analogous in their reactions to Chlorine, Iodine, Bromine, etc.

In a salt formed by an Oxygen Acid, both the basic and acid elements appear. Thus the common *Nitre*, which is a Nitrate of Potash, is found by analysis to contain Oxide of Potassium as a base, in a state of combination with Nitric Acid. But if a salt be formed by neutralizing an alkali with *a Hydrogen Acid*, the product in that case does not contain all the elements. This is seen from the following example:—

Hydrochloric Acid + Soda
= Chloride of Sodium + Water;

or, stated more at length,—

(Chlorine Hydrogen) + (Oxygen Sodium)
= (Chlorine Sodium)   + (Oxygen Hydrogen).

Observe that the Hydrogen and Oxygen, being present in the correct proportions, unite to form Water, which is an Oxide of Hydrogen. This water passes off when the solution is evaporated, and leaves the dry crystals of salt. On the other hand, with the Oxyacid Salts, the elementary Hydrogen being absent, no water is formed, and the Oxygen remains.

It must therefore be borne in mind that salts like the Chlorides, Bromides, Iodides, etc. contain only *two* elements; but that in the Oxyacid Salts, such as Sulphates, Nitrates, Acetates, *three* are present. Thus, Nitrate of Silver consists of Nitrogen, Oxygen, and Silver, but Chloride of Silver contains simply Chlorine and metallic Silver united, without Oxygen.

The Hydracid salts however, when decomposed, yield products similar to the Oxyacid salts. For instance, if Iodide of Potassium be dissolved in water, and dilute Sulphuric Acid added, this acid, being powerful in its chemical affinities, tends to appropriate to itself the alkali; but it does not remove *Potassium* and liberate *Iodine*, but takes the *Oxide* of Potassium and sets free *Hydriodic Acid.* In other words, as an atom of water is produced during the *formation* of a Hydracid Salt, so is an atom destroyed and made to yield up its elements in the *decomposition* of a Hydracid Salt.

The reaction of dilute Sulphuric Acid upon Iodide of Potassium may be stated thus :—

    Sulphuric Acid *plus* (Iodine Potassium) *plus* (Hydrogen Oxygen)
equals (Sulphuric Acid, Oxygen Potassium) or Sulphate of Potash,
    and (Hydrogen Iodine)   or Hydriodic Acid.

THE NATURE OF CHEMICAL AFFINITY FURTHER ILLUSTRATED.

Illustration from the Non-metallic Elements.—If a stream

of Chlorine gas be passed into a solution containing the
same salt as before mentioned, viz. the Iodide of Potassium, the result is to liberate a certain portion of Iodine,
which dissolves in the liquid, and tinges it of a brown
colour. The element Chlorine, possessing a degree of chemical energy superior to that of Iodine, prevails over it,
and removes the Potassium with which the Iodine was
previously combined.

Chlorine + Iodide of Potassium
= Iodine + Chloride of Potassium.

The same Law illustrated by the Metals.—A strip of Iron
dipped in solution of Nitrate of Silver becomes immediately coated with metallic Silver; but a piece of Silver-
foil may be left for any length of time in Sulphate of Iron
without undergoing change : the difference depends upon
the fact, that metallic Iron has a greater attraction for
Oxygen than Silver, and hence it displaces it from its solution.

Iron + Nitrate of Silver
= Silver + Nitrate of Iron.

Illustrations amongst Binary Compounds.—If a few
drops of solution of Potash be added to solution of Nitrate
of Silver, a brown deposit is formed, which is the Oxide
of Silver, sparingly soluble in water. That is to say, as
a stronger *metal* displaces *metallic Silver,* so does an *oxide*
of the same metal displace *Oxide of Silver.* Therefore
bases like the alkalies, alkaline earths, etc. cannot exist in
a free state in solutions of the salts of weaker bases,—a
liquid containing Nitrate of Silver could not also contain
free Potash or Ammonia.

In the list given at page 306, the metallic elements are
arranged principally in the order of their chemical affinities; those of Potassium, Sodium, Barium, etc. being the
most marked.

As the alkalies displace the weaker bases from their

combination with acids, so the strong *acids* displace weak
acids from their combination with bases. Thus, as

$$\text{Oxide of Potassium} + \text{Acetate of Silver}$$
$$= \text{Oxide of Silver} \quad + \text{Acetate of Potash};$$

So

$$\text{Nitric Acid} + \text{Acetate of Silver}$$
$$= \text{Acetic Acid} + \text{Nitrate of Silver}.$$

In the list of acids, Sulphuric Acid is usually placed
first as being the strongest, and Carbonic Acid, which is
a gaseous substance, last. The vegetable acids, such as
Acetic, Tartaric, etc., are *intermediate*, being weaker than
the mineral acids, but stronger than Carbonic, or Hydro-
cyanic Acid.

*The order of decompositions affected by the insolubility
or the volatility of the products which may be formed.*—It
might be inferred from remarks already made, that on
mixing saline solutions, a gradual interchange of elements
would take place, until the strongest acids were associated
with the strongest bases, and *vice versâ.* There are many
causes however which interfere to prevent this; one of
which is *volatility.*—

The violent effervescence which takes place on treating
a *Carbonate* of any kind with an acid is due to the *gaseous*
nature of Carbonic Acid and its escape in that form, which
greatly facilitates the decomposition.

Insolubility is also a cause which exercises a great influence on the result which will follow in mixing solutions.
If the formation of an insoluble substance is possible by
any interchange of elements, it will take place. A solution of Chloride of Sodium added to Nitrate of Silver
invariably produces Chloride of Silver; the *insolubility* of
Chloride of Silver being the cause which determines its
formation.

So again, Sulphate of Lead and Protonitrate of Iron
are produced by mixing Nitrate of Lead with Sulphate of
Iron; but if Nitrate of *Potash* be substituted for Nitrate

of Lead, the result is uncertain, because there are no elements present which can, by interchanging, form an insoluble salt; Sulphate of Potash, although *sparingly* soluble in water, not being *insoluble*, like the Sulphate of Lead or the Sulphate of Baryta.

ON CHEMICAL NOMENCLATURE.

The nomenclature of the chemical *elements* is mostly independent of any rule; but an attempt has been made to obviate this in the case of those of later discovery. Thus the names of the newly-found *metals* usually end in *um*, as Potassium, Sodium, Barium, Calcium, etc.; and those elements which possess analogous characters have corresponding terminations assigned to them, as Chlorine, Bromine, Iodine, Fluorine, etc.

Nomenclature of Binary Compounds.—These are often named by attaching the termination *ide* to the more important element of the two; as, the Ox*ide* of Hydrogen, or Water; the Chlor*ide* of Silver; the Sulph*ide* of Silver. Binary compounds of Sulphur however are sometimes termed Sulphurets, as the *Sulphuret* or the *Sulphide* of Silver indifferently.

When the same body combines with Oxygen, or the corresponding element, in more than one proportion, the prefix *proto* is applied to that containing the least Oxygen; *sesqui* to that with once and a half as much as the *proto*; *bi* or *bin* to that with twice as much; and *per* to the one containing the most Oxygen of all. As examples, take the following:—The Protoxide of Iron; the Sesquioxide of Iron: the Protochloride of Mercury; the Bichloride of Mercury. In these examples the Sesquioxide of Iron is also a *Per*oxide, because no higher simple oxide is known, and the Bichloride of Mercury is a *Per*chloride for a similar reason.

When an inferior compound is discovered, it is often termed *sub*; as the Suboxide of Silver, the Subchloride of Silver. These bodies contain the least known quantity of Oxygen and Chlorine respectively, and are hence entitled to the prefix *proto;* but being of minor importance, they are excepted from the general rule.

The combinations of metallic elements with each other are termed "alloys;" or if containing Mercury, "amalgams."

Nomenclature of Binary Compounds possessing acid properties.—These are named on a different principle. The termination *ic* is applied to one element. Thus, taking as an illustration the liquid known as "Oil of Vitriol," it is truly an *Oxide* of Sulphur, but as it possesses strong acid properties it is termed Sulphur*ic Acid.* So Nitric Acid is an Oxide of Nitrogen; Carbonic Acid is an Oxide of Carbon, etc. When there are two oxides of the same element, both possessing acid properties, the most important has the termination *ic,* and the other *ous;* as Sulphuric Acid, Sulphur*ous* Acid; Nitric Acid, Nitr*ous* Acid.

Nomenclature of the Hydracids.—The Hydrogen Acids are distinguished from Oxyacids by retaining the names of both constituents, the termination *ic* being annexed as usual. Thus, *Hydro*chloric Acid, or the Chloride of Hydrogen; *Hydr*iodic Acid, or the Iodide of Hydrogen.

Further illustrations of the nomenclature of Binary Compounds.—The Oxides of Nitrogen, and also of Sulphur, afford an interesting illustration of the principles of nomenclature. The former are as follows:—

	Nitrogen.	Oxygen.
Protoxide of Nitrogen . . .	1 atom.	1 atom.
Binoxide of Nitrogen . . .	1 ,,	2 ,,
Nitrous Acid	1 ,,	3 ,,
Peroxide of Nitrogen . . .	1 ,,	4 ,,
Nitric Acid	1 ,,	5 ,,

Observe, that two only out of the five possess acid properties, the others being simple oxides. Nitric Acid is, strictly speaking, the "Peroxide," but as it belongs to the class of acids, that term naturally falls to the compound below.

The binary compounds of Sulphur with Oxygen all possess acid properties; they may be represented (in part) as follows:—

	Sulphur.	Oxygen.
Hyposulphurous Acid . . .	2 atoms.	2 atoms.
Sulphurous Acid	1 „	2 „
Hyposulphuric Acid . . .	2 „	5 „
Sulphuric Acid	1 „	3 „

In this case the Sulphuric and Sulphurous Acids had become familiarly known before the others, intermediate in composition, were discovered. Hence, to avoid the confusion which would result from changing the nomenclature, the new bodies are termed *Hypo*sulphuric and *Hypo*sulphurous (from ὑπο, *under*).

Nomenclature of Salts.—Salts are named according to the acid they contain; the termination *ic* being changed into *ate*, and *ous* into *ite*. Thus, Sulphuric Acid forms Sulph*ates*; Nitric Acid, Nitr*ates*; but Sulphur*ous* Acid forms Sulph*ites*, and Nitrous Acid, Nitr*ites*.

In naming a salt, the base is always placed *after* the acid, the term *oxide* being omitted; thus, *Nitrate of Oxide of Silver* is more shortly known as "Nitrate of Silver," the presence of Oxygen being understood.

When there are two oxides of the same base, both of which are *salifiable*,—in naming the salts, the term *proto* is prefixed to the acid of the salt formed by the lowest, and *per* to that of the higher oxide; as, the *Proto*sulphate of Iron, or Sulphate of the Protoxide; the *Per*sulphate of Iron, or Sulphate of the Peroxide.

Many salts contain more than one atom of acid to each atom of base. In that case, the usual prefixes expressive of quantity are adopted: thus, the *Bi*sulphate of Potash contains twice as much Sulphuric Acid as the neutral Sulphate, etc.

On the other hand, there are salts in which the base is in excess with regard to the acid, and which are usually known as "basic salts;" thus, the red powder which deposits from solution of Sulphate of Iron, is a *basic* Per-sulphate of Iron, or a Sulphate of the Peroxide of Iron with more than the normal proportion of oxide.

Nomenclature of the Hydracid Salts.—The composition of these salts being different from those formed by Oxygen Acids, the nomenclature varies also. Thus, in neutralizing Hydrochloric Acid with Soda, the product formed is not known as Hydrochlorate of Soda, but as *Chloride of Sodium;* this salt, and others of a similar constitution, being *binary*, and not *ternary*, compounds. The salt produced by Hydrochloric Acid and *Ammonia* however is often called " Muriate or Hydrochlorate of Ammonia," although more strictly it should be the *Chloride of Ammonium.*

ON SYMBOLIC NOTATION.

The list of symbols employed to represent the various elementary bodies is given at page 306.—Commonly the initial letter of the Latin name is used, a second or smaller letter being added when two elements correspond in their initials : thus C stands for Carbon, Cl for Chlorine, Cd for Cadmium, and Cu for Copper.

The chemical symbol however does not simply represent a particular element; it denotes also a definite weight, or equivalent proportion, of that element. This will be explained more fully in the succeeding pages, when speaking of the Laws of Combination.

Formulæ of Compounds.—In the *nomenclature* of compounds it is usual to place the Oxygen or analogous element *first* in the case of binary compounds, and the acid before the base in the ternary compounds, or salts; but in representing them *symbolically* this order is reversed : thus, Oxide of Silver is written AgO, and never as OAg; Nitrate of Silver as $AgO\ NO_5$, not NO_5AgO.

The juxtaposition of symbols expresses combination; thus, FeO is a compound of one proportion of Iron with one of Oxygen, or the " Protoxide of Iron." If more than one equivalent be present, small figures are placed below the symbols: thus, Fe_2O_3 represents two equivalents of Iron united with three of Oxygen, or the "Peroxide of Iron;" SO_3, one equivalent of Sulphur with three of Oxygen, or Sulphuric Acid.

Larger figures placed before and in the same line with the symbols, affect the *whole compound* which the symbols express: thus, $2 SO_3$ means two equivalents of Sulphuric Acid; $3 NO_5$, three equivalents of Nitric Acid. The interposition of a comma prevents the influence of the large figure from extending further. Thus, the double Hyposulphite of Soda and Silver is represented as follows :—

$$2 NaO S_2O_2, AgO S_2O_2,$$

or *two* equivalents of Hyposulphite of Soda with *one* of Hyposulphite of Silver; the large figure referring only to the first half of the formula. Sometimes brackets, etc. are employed, in order to render a complicated formula more plain. For example, the formula for the double Hyposulphite of Gold and Soda, or "Sel d'or," may be written thus :—

$$3 (NaO S_2O_2) AuO S_2O_2 + 4 HO.$$

In this formula, the *plus* sign ($+$) denotes that the four atoms of water which follow, are less intimately united with the framework of the salt than the other constituents.

The use of a *plus* sign is commonly adopted in representing salts which contain water of crystallization. Thus, the formula for the crystallized Protosulphate of Iron is written as follows :—

$$FeO SO_3 + 7 HO.$$

These atoms of water are driven off by the application of heat, leaving a white substance, which is the Anhydrous salt, and would be written simply as $FeO SO_3$.

The *plus* sign however is often employed in token of simple *addition*, no combination of any kind being intended. Thus the decomposition which follows on mixing

Chloride of Sodium with Nitrate of Silver may be written as follows :—

$$NaCl + AgO\ NO_5 = AgCl + NaO\ NO_5 ;$$

that is,—

Chloride of Sodium *added* *to* Nitrate of Silver.
= Chloride of Silver *and* Nitrate of Soda.

ON EQUIVALENT PROPORTIONS.

When elementary or compound bodies enter into chemical union with each other, they do not combine in indefinite proportions, as in the case of a mixture of two liquids, or the solution of a saline body in water. On the other hand, a certain definite weight of the one unites with an equally definite weight of the other; and if an excess of either be present, it remains free and uncombined.

Thus, if we take a *single grain* of the element Hydrogen —to convert that grain into Water there will be required exactly 8 grains of Oxygen ; and if a larger quantity than this were added, as for instance *ten grains*, then two grains would be over and above. So, to form *Hydrochloric Acid*, 1 grain of Hydrogen takes 36 grains of Chlorine :—for the *Hydriodic Acid*, 1 grain of Hydrogen unites with 126 grains of Iodine.

Again, if separate portions of metallic Silver, of 108 grains each, are weighed out,—in order to convert them into Oxide, Chloride, and Iodide of Silver respectively, there would be required

Oxygen . . . . 8 grains.
Chlorine . . . . 36 „
Iodine . . . . 126 „

Therefore it appears that 8 grains of Oxygen are *equivalent* to 36 grains of Chlorine and to 126 grains of Iodine, seeing that these quantities all play the same part in combining ; and so it is with regard to the other elements,—to every one of them a figure can be assigned which represents the number of parts by weight in which that element unites with others. These figures are the " equivalents " or " combining proportions," and they are denoted by the *symbol* of the element. A symbol does not stand as a simple representative of an element, but as a representative of *one equivalent* of an element. Thus " O " indicates 8 parts by weight of Oxygen; " Cl " one equivalent, or 36 parts by weight, of Chlorine; and so with the rest.

Observe however that these figures, termed " equivalents," do not refer to the *actual number* of parts by weight, but only to the *ratio* which exists between them: if Oxygen is 8, then Chlorine is 36; but if we term Oxygen 100, as some have proposed, then Chlorine would be 442·65.

In the scale of equivalents now usually adopted, Hydrogen, as being the lowest of all, is taken as unity, and the others are related to it.

Equivalents of Compounds.—The law of equivalent proportions applies to compounds as well as to simple bodies, the combining proportion of a compound being always the *sum* of the equivalents of its constituents. Thus Sulphur is 16, and Oxygen 8, therefore Sulphuric Acid, or SO_3, equals 40. The equivalent of Nitrogen is 14, that of Nitric Acid, or NO_5, is 54.

The same rule applies with regard to salts. Take for instance the Nitrate of Silver: it contains

	Equivalent.
Nitrogen	14
6 Oxygen	48
Silver	108

Total of equivalents, or equivalent of the Nitrate of Silver } . 170

Practical application of the Laws of Combination.—The utility of being acquainted with the law of combining proportions is obvious when their nature is understood. As bodies both unite with and replace each other in equiva-

Y

lents, a simple calculation shows at once how much of each element or compound will be required in a given reaction. Thus, supposing it be desired to convert 100 grains of Nitrate of Silver into *Chloride* of Silver, the weight of Chloride of Sodium which will be necessary is deduced thus:—one equivalent, or 170 parts, of Nitrate of Silver, is decomposed by an equivalent, or 60 parts, of Chloride of Sodium. Therefore

$$\text{as } 170 : 60 :: 100 : 35\cdot2 ;$$

that is, 35·2 grains of Salt will precipitate, in the state of Chloride, the whole of the Silver contained in 100 grains of Nitrate.

So again, in order to form the Iodide of Silver, the proportions in which the two salts should be mixed is thus shown. The equivalent of Iodide of Potassium is 166, and that of Nitrate of Silver is 170. These numbers so nearly correspond, that it is common to direct that *equal weights* of the two salts should be taken.

One more illustration will suffice. Supposing it be required to form 20 grains of Iodide of Silver—how much Iodide of Potassium and Nitrate of Silver must be used? One equivalent, or 166 parts, of Iodide of Potassium, will yield an equivalent, or 234 parts, of Iodide of Silver; therefore

$$\text{as } 234 : 166 :: 20 : 14\cdot2.$$

Hence, if 14·2 grains of the Iodide of Potassium be dissolved in water, and an equivalent quantity, viz. 14·5 grains, of the Nitrate of Silver added, the yellow precipitate, when washed and dried, will weigh precisely 20 grains.

ON THE ATOMIC THEORY.

The atomic theory, originally proposed by Dalton, so much facilitates the comprehension of chemical reactions generally, that it may be useful to give a short sketch of it.

It is supposed that all matter is made up of an infinite number of minute atoms, which are elementary, and do not admit of further division. Each of these atoms possesses an actual weight, although inappreciable by our present methods of investigation. Simple atoms, by uniting with each other, form *compound atoms;* and when these compounds are broken up, the elementary constituent atoms are not destroyed, but separate from each other, in possession of all their original properties.

In representing the simple atomic structure of bodies, circles may be used, as in the following diagram.

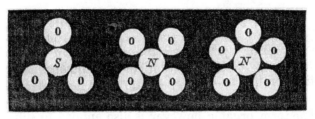

Fig. 1.   Fig. 2.   Fig. 3.

Fig. 1 is a compound atom of Sulphuric Acid, consisting of an atom of Sulphur united intimately with three of Oxygen; fig. 2 is an atom of Peroxide of Nitrogen, NO_4; and fig. 3, an atom of Nitric Acid, composed of Nitrogen 1 atom, Oxygen 5 atoms, or in symbols NO_5.

The term "atomic weight" substituted for equivalent proportion.—If we suppose that the simple atoms of different kinds of matter *differ in weight,* and that this difference is expressed by their equivalent numbers, the whole laws of combination follow by the simplest reasoning. It is easy to understand that an atom of one element, or compound, would displace, or be substituted for, a single atom of another; therefore, taking as the illustration the decomposition of Iodide of Potassium by Chlorine,—the weight of the latter element required to liberate 126 grains of Iodine is 36 grains, *because the weights of the atoms of those two elementary bodies are as 36 to 126.* So again, in the reaction between Chloride of Sodium and Nitrate of Silver, a compound atom of the former, represented by the weight 60, reacts upon a compound atom of the latter, which equals 170.

Therefore in place of the term "equivalent" or "combining proportion," it is more usual to employ that of "atomic weight." Thus the atomic weight of Oxygen is 8, represented by the symbol O; that of Sulphur is 16; hence the atomic weight of the compound atom of Sulphuric Acid, or SO_3, is necessarily equal to the combined weights of the four simple atoms; *id est*, $16+24=40$.

ON THE CHEMISTRY OF ORGANIC SUBSTANCES.

By "organic" substances are meant those which have possessed *life*, with definite organs and tissues, in contradistinction to the various forms of dead inorganic matter, in which no structural organization of that kind is found.

The term organic however is also applied to substances which are obtained by chemical processes from the vegetable and animal kingdoms, although they cannot themselves be said to be living bodies; thus Acetic Acid, procured by the distillation of woody fibre, and Alcohol, by fermentation from sugar, are strictly organic substances.

The class of organic bodies embraces a great variety of products; which, like inorganic Oxides, may be divided into neutral, acid, and basic.

The organic *acids* are numerous, including Acetic Acid, Tartaric, Citric, and a variety of others.

The *neutral substances* cannot easily be assimilated to any class of inorganic compounds; as examples, take Starch, Sugar, Lignine, etc.

The *bases* are also a large class. They are mostly rare substances, not familiarly known: Morphia, obtained from Opium; Quinia, from Quinine; Nicotine, from Tobacco, are illustrations.

Composition of organic and inorganic bodies contrasted.
—There are more than fifty elementary substances found
in the inorganic kingdom, but only *four*, commonly speaking, in the organic : these four are Carbon, Hydrogen, Nitrogen, and Oxygen.

Some organic bodies,—oil of turpentine, naphtha, etc.,
contain only Carbon and Hydrogen;. many others, such as
sugar, gum, alcohol, fats, vegetable acids—Carbon, Hydrogen, and Oxygen. The *Nitrogenous bodies*, so called, containing Nitrogen in addition to the other elements, are
principally substances derived from animal and vegetable
tissues, such as Albumen, Caseine, Gelatine, etc.; Sulphur
and Phosphorus are also present in many of the Nitrogenous bodies, but only to a small extent.

Organic substances, although simple as regards the
number of elements involved in their formation, are often
highly complex in the arrangement of the atoms ; this may
be illustrated by the following formulæ :—

Starch	$C_{24}H_{20}O_{20}$
Lignine	$C_{24}H_{20}O_{20}$
Cane Sugar	$C_{24}H_{22}O_{22}$
Grape Sugar . . .	$C_{24}H_{28}O_{28}$

Inorganic bodies, as already shown, unite *in pairs*,—two
elements join to form a binary compound ; two binary compounds produce a salt ; two salts associated together form
a double salt. With organic bodies however the arrangement is different,—the elementary atoms are all grouped
equally in one compound atom, which is highly complex in
structure, and cannot be split up into binary products.

Observe also, as characteristic of Organic Chemistry,
the apparent similarity in composition between bodies
which differ widely in properties. As examples take *Lignine*, or cotton fibre, and *Starch*,—each of which contains
the three elements united as $C_{24}H_{20}O_{20}$.

*Mode of distinguishing between Organic and Inorganic
matter.*—A simple means of doing this is as follows :—

place the suspected substance upon a piece of Platinum-foil, and heat it to redness with a spirit-lamp : if it first *blackens*, and then burns completely away, it is probably of organic origin. This test depends upon the fact, that the constituent elements of organic bodies are all either themselves volatile, or capable of forming volatile combinations with Oxygen. Inorganic substances, on the other hand, are often unaffected by heat, or, if volatile, are dissipated without previous charring.

The action of heat upon organic matter may further be illustrated by the combustion of coal or wood in an ordinary furnace ;—first, an escape of Carbon and Hydrogen, united in the form of volatile gaseous matter, takes place, leaving behind a black cinder, which consists of Carbon and inorganic matter combined; afterwards this Carbon burns away into Carbonic Acid, and a grey ash is left which is composed of inorganic salts, and is indestructible by heat.

CHAPTER II.

VOCABULARY OF PHOTOGRAPHIC CHEMICALS.

ACETIC ACID.

Symbol, $C_4H_3O_3 + HO$. Atomic weight, 60.

ACETIC Acid is a product of the *oxidation* of Alcohol. Spirituous liquids, when perfectly pure, are not affected by exposure to air; but if a portion of yeast, or Nitrogenous organic matter of any kind, be added, it soon acts as a *ferment*, and causes the spirit to unite with oxygen derived from the atmosphere, and so to become *sour* from formation of Acetic Acid, or "vinegar."

Acetic Acid is also produced on a large scale by heating *wood* in close vessels: a substance distils over which is Acetic Acid contaminated with empyreumatic and tarry matter; it is termed Pyroligneous Acid, and is much used in commerce.

The most concentrated Acetic Acid may be obtained by neutralizing common vinegar with Carbonate of Soda, and crystallizing out the Acetate of Soda so formed; this Acetate of Soda is then distilled with Sulphuric Acid, which removes the Soda and liberates Acetic Acid: the Acetic Acid being volatile, distils over, and may be condensed.

Properties of Acetic Acid.—The strongest acid contains only a single atom of water; it is sold under the name

of "Glacial Acetic Acid," so called from its property of
solidifying at a moderately low temperature. At about
50° the crystals melt, and form a limpid liquid of pungent odour and a density nearly corresponding to that of
water; the specific gravity of Acetic Acid however is no
test of its real strength, which can only be estimated by
analysis.

The commercial *Glacial* Acetic Acid is often diluted
with water, which may be suspected if it does not solidify
during the cold winter months. Sulphurous and Hydrochloric Acids are also common impurities. They are injurious in Photographic Processes, from their property of
precipitating Nitrate of Silver. To detect them proceed
as follows:—dissolve a small crystal of Nitrate of Silver in
a few drops of water, and add to it about half a drachm of
the Glacial Acid; the mixture should remain quite clear
even when exposed to the light. Hydrochloric and Sulphurous Acid produce a white deposit of Chloride or Sulphite of Silver; and if *Aldehyde* or volatile tarry matter
be present in the Acetic Acid, the mixture with Nitrate
of Silver, although clear at first, becomes discoloured by
the action of light.

Glacial Acetic Acid sometimes has a smell of garlic. In
this state it probably contains an organic Sulphur Acid,
and is unfit for use.

Many employ a cheaper form of Acetic Acid, sold
by druggists as "Beaufoy's" acid; it should be of the
strength of the Acetic Acid fortiss. of the London Pharmacopœia, containing 30 per cent. real acid. It will be
advisable to test it for Sulphuric Acid (see Sulphuric
Acid), and other impurities, before use.

ACETATE OF SILVER. *See* SILVER, ACETATE OF.

ALBUMEN.

Albumen is an organic principle found both in the
animal and vegetable kingdom. Its properties are best

studied in the *white of egg*, which is a very pure form of Albumen.

Albumen is capable of existing in two states; in one of which it is soluble, in the other insoluble, in water. The aqueous solution of the soluble variety gives a slightly alkaline reaction to test-paper; it is somewhat thick and glutinous, but becomes more fluid on the addition of a small quantity of an alkali, such as Potash or Ammonia.

Soluble Albumen may be converted into the *insoluble* form in the following ways:—

1. *By the application of heat.*— A moderately strong solution of Albumen becomes opalescent and coagulates on being heated to about 150° Fahrenheit, but a temperature of 212° is required if the liquid is very dilute. A layer of *dried* Albumen cannot easily be coagulated by the mere application of heat.

2. *By addition of strong acids.*—Nitric Acid coagulates Albumen perfectly without the aid of heat. Acetic Acid however acts differently, appearing to enter into combination with the Albumen, and forming a compound soluble in warm water acidified by Acetic Acid.

3. *By the action of metallic salts.*—Many of the salts of the metals coagulate Albumen completely. Nitrate of Silver does so; also the Bichloride of Mercury. Ammoniacal Oxide of Silver however does not coagulate Albumen.

The white precipitate formed on mixing Albumen with Nitrate of Silver is a chemical compound of the animal matter with Protoxide of Silver. This substance, which has been termed Albuminate of Silver, is soluble in Ammonia and Hyposulphite of Soda; but after exposure to light, or heating in a current of Hydrogen gas, it assumes a brick-red colour, being probably reduced to the condition of an organic compound of a *Suboxide* of Silver. It is then almost insoluble in Ammonia, but enough dissolves to tinge the liquid wine-red. The *red coloration* of solution of Nitrate of Silver employed in sensitizing the

Albuminized photographic paper is probably produced by the same compound, although often referred to the presence of Sulphuret of Silver.

Albumen also combines with Lime and Baryta. When Chloride of Barium is used with Albumen, a white precipitate of this kind usually forms.

Chemical composition of Albumen.—Albumen belongs to the *Nitrogenous* class of organic substances (see page 325). It also contains small quantities of Sulphur and Phosphorus.

ALCOHOL.

Symbol, $C_4H_6O_2$. Atomic weight, 46.

Alcohol is obtained by the careful distillation of any spirituous or fermented liquor. If wine or beer be placed in a retort, and heat applied, the Alcohol, being more volatile than water, rises first, and is condensed in an appropriate receiver; a portion of the vapour of water however passes over with the Alcohol, and dilutes it to a certain extent, forming what is termed " Spirits of Wine." Much of this water may be removed by redistillation from Carbonate of Potash, in the manner described at page 196 of this work; but in order to render the Alcohol thoroughly *anhydrous*, it is necessary to employ *quicklime*, which possesses a still greater attraction for water. An equal weight of this powdered lime is mixed with strong Alcohol of ·823, and the two are distilled together.

Properties of Alcohol.—Pure anhydrous Alcohol is a limpid liquid, of an agreeable odour and pungent taste; sp. gr. at 60°, ·794. It absorbs vapour of water, and becomes diluted by exposure to damp air; boils at 173° Fahr. It has never been frozen.

Alcohol distilled from Carbonate of Potash has a specific gravity of ·815 to ·823, and contains 90 to 93 per cent. of real spirit.

The specific gravity of ordinary rectified Spirits of Wine

is usually about ·840, and it contains 80 to 83 per cent. of absolute Alcohol.

AMMONIA.

Symbol, NH_3 or NH_4O. Atomic weight, 17.

The liquid known by this name is an aqueous solution of the volatile gas Ammonia. Ammoniacal gas contains one atom of Nitrogen combined with three of Hydrogen: these two elementary bodies exhibit no affinity for each other, but they can be made to unite under certain circumstances, and the result is Ammonia.

Properties of Ammonia.—Ammoniacal gas is soluble in water to a large extent; the solution possessing those properties which are termed alkaline (see page 308). Ammonia however differs from the other alkalies in one important particular—it is volatile: hence the original colour of turmeric-paper affected by Ammonia is restored on the application of heat. Solution of Ammonia absorbs Carbonic Acid rapidly from the air, and is converted into Carbonate of Ammonia; it should therefore be preserved in stoppered bottles. Besides Carbonate, commercial Ammonia often contains Chloride of Ammonium, recognized by the white precipitate given by Nitrate of Silver after acidifying with pure Nitric Acid.

The strength of commercial Ammonia varies greatly; that sold for pharmaceutical purposes under the name of Liquor Ammoniæ, contains about 10 per cent. of real Ammonia. The sp. gr. of aqueous Ammonia *diminishes* with the proportion of Ammonia present, the Liquor Ammoniæ being usually about ·936.

Ammonia, although forming a large class of salts, appears at first sight to contrast strongly in composition with the alkalies proper, such as Potash and Soda. Mineral bases generally are *protoxides of metals*, as already shown at page 308, but Ammonia consists simply of Nitrogen and Hydrogen united without Oxygen. The following remarks may perhaps tend somewhat to elucidate the difficulty:—

Theory of Ammonium.—This theory supposes the existence of a substance possessing the properties of a *metal*, but differing from metallic bodies generally in being *compound* in structure: the formula assigned to it is NH_4, one atom of Nitrogen united with four of Hydrogen. This hypothetical metal is termed "Ammonium;" and Ammonia, associated with an atom of water, may be viewed as its *Oxide*, for $NH_3 + HO$ plainly equals NH_4O. Thus, as Potash is the Oxide of *Potassium*, so Ammonia is the Oxide of *Ammonium*.

The composition of the *salts* of Ammonia is on this view assimilated to those of the alkalies proper. Thus, Sulphate of Ammonia is a Sulphate of the Oxide of Ammonium; Muriate or Hydrochlorate of Ammonia is a Chloride of Ammonium, etc.

AMMONIO-NITRATE OF SILVER.

See SILVER, AMMONIO-NITRATE OF.

AQUA REGIA. *See* NITRO-HYDROCHLORIC ACID.

BARYTA, NITRATE OF. *See* NITRATE OF BARYTA.

BICHLORIDE OF MERCURY.

See MERCURY, BICHLORIDE OF.

BROMINE.

Symbol, Br. Atomic weight, 78.

This elementary substance is obtained from the uncrystallizable residuum of sea-water, termed *bittern*. It exists in the water in very minute proportion, combined with Magnesium in the form of a soluble Bromide of Magnesium.

Properties.—Bromine is a deep reddish-brown liquid of

a disagreeable odour, and fuming strongly at common temperatures; sparingly soluble in water (1 part in 23, Löwig), but more abundantly so in Alcohol, and especially in Ether. It is very heavy, having a specific gravity of 3·0.

Bromine is closely analogous to Chlorine and Iodine in its chemical properties. It stands on the list intermediately between the two; its affinities being stronger than those of Iodine, but weaker than Chlorine (see Chlorine).

It forms a large class of salts, of which the Bromides of Potassium, Cadmium, and Silver are the most familiar to Photographers.

BROMIDE OF POTASSIUM.

Symbol, KBr. Atomic weight, 118.

Bromide of Potassium is prepared by adding Bromine to Caustic Potash, and heating the product, which is a mixture of Bromide of Potassium and Bromate of Potash, to redness, in order to drive off the Oxygen from the latter salt. It crystallizes in anhydrous cubes, like the Chloride and Iodide of Potassium; it is easily soluble in water, but more sparingly so in Alcohol; it yields red fumes of Bromine when acted upon by Sulphuric Acid.

BROMIDE OF SILVER. *See* SILVER, BROMIDE OF.

CARBONATE OF SODA.

Symbol, $NaO\ Co_2 + 10\ Aq$.

This salt was formerly obtained from the ashes of seaweeds, but is now more economically manufactured on a large scale from common salt. The Chloride of Sodium is first converted into Sulphate of Soda, and afterwards the Sulphate into Carbonate of Soda.

Properties.—The perfect crystals contain ten atoms of water, which are driven off by the application of heat, leaving a white powder—the anhydrous Carbonate. *Common*

Washing Soda is a neutral Carbonate, contaminated to a certain extent with Chloride of Sodium and Sulphate of Soda. The Carbonate used for effervescing draughts is either a Bicarbonate with 1 atom of water, or a Sesquicarbonate, containing about 40 per cent. of real alkali; it is therefore nearly double as strong as the washing Carbonate, which contains about 22 per cent. of Soda. Carbonate of Soda is soluble in twice its weight of water at 60°, the solution being strongly alkaline.

CARBONATE OF POTASH.

See POTASH, CARBONATE OF.

CASEINE. *See* MILK.

CHARCOAL, ANIMAL.

Animal Charcoal is obtained by heating animal substances, such as bones, dried blood, horns, etc., to redness, in close vessels, until all volatile empyreumatic matters have been driven off, and a residue of Carbon remains. When prepared from bones it contains a large quantity of inorganic matter in the shape of Carbonate and Phosphate of Lime, the former of which produces *alkalinity* in reacting upon Nitrate of Silver (see p. 89). Animal Charcoal is freed from these earthy salts by repeated digestion in Hydrochloric Acid; but unless very carefully washed it is apt to retain an acid reaction, and so to liberate free Nitric Acid when added to solution of Nitrate of Silver.

Properties.—Animal Charcoal, when pure, consists solely of Carbon, and burns away in the air without leaving any residue: it is remarkable for its property of decolorizing solutions; the organic colouring substance being separated, but not actually *destroyed*, as it is by *Chlorine* employed as a bleaching agent. This power of absorbing colouring matter is not possessed in an equal degree by all varieties of Charcoal, but is in great measure peculiar to those derived from the animal kingdom.

CHINA CLAY, OR KAOLIN.

This is prepared, by careful levigation, from mouldering granite and other disintegrated felspathic rocks. It consists of the *Silicate of Alumina,*—that is, of Silicic Acid or *Flint,* which is an Oxide of Silicon, united with the base Alumina (Oxide of Aluminum). Kaolin is perfectly insoluble in water and acids, and produces no decomposition in solution of Nitrate of Silver. It is employed by Photographers to decolorize solutions of Nitrate of Silver which have become brown from the action of Albumen or other organic matters.

Commercial Kaolin may contain chalk, in which state it produces alkalinity in solution of Nitrate of Silver. The impurity, detected by its effervescence with acids, is removed by washing the Kaolin in diluted vinegar and subsequently in water.

CHLORINE.

Symbol, Cl. Atomic weight, 36.

Chlorine is a chemical element found abundantly in nature, combined with metallic Sodium in the form of Chloride of Sodium, or Sea-salt.

Preparation.—By distilling common Salt with Sulphuric Acid, Sulphate of Soda and Hydrochloric Acid are formed. Hydrochloric Acid contains Chlorine combined with Hydrogen; by the action of *nascent* Oxygen (see Oxygen), the Hydrogen may be removed in the form of water, and the Chlorine left alone.

Properties.—Chlorine is a greenish-yellow gas, of a pungent and suffocating odour; soluble to a considerable extent in water, the solution possessing the odour and colour of the gas. It is nearly $2\frac{1}{2}$ times as heavy as a corresponding bulk of atmospheric air.

Chemical properties.—Chlorine belongs to a small natural group of elements which contains also Bromine, Iodine, and Fluorine. They are characterized by having a strong affinity for Hydrogen, and also for the metals; but are comparatively indifferent to Oxygen. Many metallic substances actually undergo *combustion* when projected into an atmosphere of Chlorine, the union between the two taking place with extreme violence. The characteristic bleaching properties of Chlorine gas are explained in the same manner :—Hydrogen is removed from the organic substance, and in that way the structure is broken up and the colour destroyed.

Chlorine is more powerful in its affinities than either Bromine or Iodine. The salts formed by these three elements are closely analogous in composition and often in properties. Those of the Alkalies, Alkaline Earths, and many of the Metals, are soluble in water; but the Silver salts are insoluble; the Lead salts sparingly so.

The combinations of Chlorine, Bromine, Iodine, and Fluorine, with Hydrogen, are acids, and neutralize Alkalies in the usual manner, with formation of Alkaline Chloride and water (see page 311).

The test by which the presence of Chlorine is detected, either free or in combination with bases, is *Nitrate of Silver;* it gives a white curdy precipitate of Chloride of Silver, insoluble in Nitric Acid, but soluble in Ammonia. The solution of Nitrate of Silver employed as the test must not contain Iodide of Silver, as this compound is precipitated by dilution.

CHLORIDE OF AMMONIUM.

Symbol, NH_4Cl. Atomic weight, 54.

This salt, also known as Muriate or Hydrochlorate of Ammonia, occurs in commerce in the form of colourless and translucent masses, which are procured by *sublimation*, the dry salt being volatile when strongly heated. It dissolves in an equal weight of boiling, or in three parts of cold water. It contains more *Chlorine* in proportion to the weight used than Chloride of Sodium, the atomic weights of the two being as 54 to 60.

CHLORIDE OF BARIUM.

Symbol, BaCl+2 HO. Atomic weight, 123.

Barium is a metallic element very closely allied to Calcium, the elementary basis of *Lime*. The Chloride of Barium is commonly employed as a test for Sulphuric Acid, with which it forms an insoluble precipitate of Sulphate of Baryta. It is also said to affect the colour of the Photographic image when used in preparing Positive paper, which may possibly be due to a chemical combination of Baryta with Albumen; but it must be remembered that this Chloride, from its high atomic weight, contains *less* Chlorine than the alkaline Chlorides (see page 124).

Properties of Chloride of Barium.—Chloride of Barium occurs in the form of white crystals, soluble in about two parts of water, at common temperature. These crystals contain two atoms of water of crystallization, which are expelled at 212°, leaving the anhydrous Chloride.

CHLORIDE OF GOLD. *See* GOLD, CHLORIDE OF.

CHLORIDE OF SODIUM.

Symbol, NaCl. Atomic weight, 60.

Common Salt exists abundantly in nature, both in the form of solid rock-salt and dissolved in the waters of the ocean.

Properties of the pure Salt.—Fusible without decomposition at low redness, but sublimes at higher temperatures; the melted salt concretes into a hard white mass on cooling. Nearly insoluble in absolute alcohol, but dissolves in

z

minute quantity in rectified spirit.   Soluble in three parts
of water, both hot and cold.   Crystallizes in cubes, which
are anhydrous.

Impurities of Common Salt.—Table Salt often contains
large quantities of the Chlorides of Magnesium and Calcium, which, being deliquescent, produce a dampness by
absorption of atmospheric moisture: Sulphate of Soda is
also commonly present.   The salt may be purified by repeated recrystallization, but it is more simple to prepare
the pure compound *directly*, by neutralizing Hydrochloric
Acid with Carbonate of Soda.

CHLORIDE OF SILVER.   *See* Silver, Chloride of.

CITRIC ACID.

This acid is found abundantly in lemon-juice and in lime-
juice.   It occurs in commerce in the form of large crystals,
which are soluble in less than their own weight of water
at 60°.

Commercial Citric Acid is sometimes mixed with Tartaric Acid.   The adulteration may be discovered by making
a concentrated solution of the acid and adding *Acetate of
Potash;* crystals of Bitartrate of Potash will separate if
Tartaric Acid be present.

Citric Acid is tribasic.   It forms with Silver a white insoluble salt, containing 3 atoms of Oxide of Silver to 1
atom of Citric Acid.   When the Citrate of Silver is heated
in a current of Hydrogen gas, a part of the acid is liberated
and the salt is reduced to a Citrate of *Suboxide* of Silver;
which is of a red colour.   The action of white light in reddening Citrate of Silver is shown by the Author to be of a
similar nature.

CYANIDE OF POTASSIUM.

Symbol, K, C_2N, or KCy.   Atomic weight, 66.

This salt is a compound of Cyanogen gas with the metal Potassium. Cyanogen is not an elementary body, like Chlorine or Iodine, but consists of Carbon and Nitrogen united in a peculiar manner. Although a compound substance, it reacts in the manner of an element, and is therefore (like *Ammonium*, previously described) an exception to the usual laws of chemistry. Many other bodies of a similar character are known.

Properties of Cyanide of Potassium.—These have been sufficiently described at page 44, to which the reader is referred.

ETHER.

Symbol, C_4H_5O. Atomic weight, 37.

Ether is obtained by distilling a mixture of Sulphuric Acid and Alcohol. If the formula of Alcohol ($C_4H_6O_2$) be compared with that of Ether, it will be seen to differ from it in the possession of an additional atom of Hydrogen and of Oxygen: in the reaction the Sulphuric Acid removes these elements in the form of *water*, and by so doing converts one atom of Alcohol into an atom of Ether. The term *Sulphuric* applied to the commercial Ether has reference only to the manner of its formation.

Properties of Ether.—The properties of Ether have been described to some extent at pages 85 and 195. The following particulars however may be added. It is neither acid nor alkaline to test-paper. Specific gravity, at 60°, about ·720. Boils at 98° Fahrenheit. The vapour is exceedingly dense, and may be seen passing off from the liquid and falling to the ground: hence the danger of pouring Ether from one bottle to another if a flame be near at hand.

Ether does not mix with water in all proportions; if the two are shaken together, after a short time the former rises and floats upon the surface. In this way a mixture of Ether and Alcohol may be purified to some extent, as in the common process of *washing* Ether. The water employed however always retains a certain portion of Ether (about a tenth part of its bulk), and acquires a strong ethereal odour; washed Ether also contains water in small quantity.

Bromine and Iodine are both soluble in Ether, and gradually react upon and decompose it.

The strong alkalies, such as Potash and Soda, also decompose Ether slightly after a time, but not immediately. Exposed to air and light, Ether is oxidized and acquires a peculiar odour (page 85).

Ether dissolves fatty and resinous substances readily, but inorganic salts are mostly insoluble in this fluid. Hence it is that Iodide of Potassium and other substances dissolved in Alcohol are precipitated to a certain extent by the addition of Ether.

FLUORIDE OF POTASSIUM.

Symbol, KF.   Atomic weight, 59.

Preparation.—Fluoride of Potassium is formed by saturating Hydrofluoric Acid with Potash, and evaporating to dryness in a platinum vessel. *Hydrofluoric Acid* contains Fluorine combined with Hydrogen; it is a powerfully acid and corrosive liquid, formed by decomposing Fluor Spar, which is a *Fluoride of Calcium*, with strong Sulphuric Acid; the action which takes place being precisely analogous to that involved in the preparation of Hydrochloric Acid.

Properties.—A deliquescent salt, occurring in small and imperfect crystals. Very soluble in water: the solution acting upon glass in the same manner as Hydrofluoric Acid.

FORMIC ACID.

Symbol, C_2HO_3.   Atomic weight, 37.

This substance was originally discovered in the *red ant* (*Formica rufa*), but it is prepared on a large scale by distilling *Starch* with Binoxide of Manganese and Sulphuric Acid.

Properties.—The strength of commercial Formic Acid is uncertain, but it is always more or less dilute. The strongest acid, as obtained by distilling Formiate of Soda with Sulphuric Acid, is a fuming liquid with a pungent odour, and containing only one atom of water. It inflames the skin in the same manner as the sting of the ant.

Formic Acid reduces the Oxides of Gold, Silver, and Mercury to the metallic state, and is itself oxidized into Carbonic Acid. The alkaline formiates also possess the same properties.

GALLIC ACID.

Symbol, $C_7H_3O_5 + H_3O$. Atomic weight, 94.

The chemistry of Gallic Acid is sufficiently described at page 27, to which the reader is referred.

GELATINE.

Symbol, $C_{13}H_{10}O_5N_2$. Atomic weight, 156.

This is an organic substance somewhat analogous to Albumen, but differing from it in properties. It is obtained by subjecting bones, hoofs, horns, calves' feet, etc., to the action of boiling water. The jelly formed on cooling is termed *size*, or, when dried and cut into slices, *glue*. Gelatine, as it is sold in the shops, is a pure form of Glue. *Isinglass* is gelatine prepared, chiefly in Russia, from the air-bladders of certain species of sturgeon.

Properties of Gelatine.—Gelatine softens and swells up in cold water, but does not *dissolve* until heated: the hot solution, on cooling, forms a tremulous jelly. One ounce of cold water will retain about three grains of Isinglass without gelatinizing; but much depends upon the temperature, a few degrees greatly affecting the result.

When long boiled in water, and especially in presence of an acid, such as the Sulphuric, Gelatine undergoes a peculiar modification, and the Solution loses either partially or entirely its property of solidifying to a jelly.

342

GLYCERINE.

Fatty bodies are resolved by treatment with an alkali into an Acid—which combines with the alkali, forming a *soap*,—and Glycerine, remaining in solution.

Pure Glycerine, as obtained by Price's patent process of distillation, is a viscid liquid of sp. gr. about 1·23; miscible in all proportions with water and Alcohol. It is peculiarly a neutral substance, exhibiting no tendency to combine with acids or bases. It has little or no action upon Nitrate of Silver in the dark, and reduces it very slowly .even when exposed to light.

GLYCYRRHIZINE.

Glycyrrhizine, obtained from the fresh root of Liquorice, is a substance intermediate in properties between a sugar and a resin. Sparingly soluble in water but very soluble in Alcohol. It precipitates strong solution of Nitrate of Silver white, but the deposit becomes reddened by exposure to light. Its preparation is described in the larger works on organic chemistry.

GOLD, CHLORIDE OF.

Symbol, $AuCl_3$. Atomic weight, 303.

This salt is formed by dissolving pure metallic Gold in Nitro-hydrochloric Acid, and evaporating at a gentle heat. The solution affords deliquescent crystals of a deep orange colour.

Chloride of Gold, in a state fit for Photographic use, may easily be obtained by the following process :—Place a half-sovereign in any convenient vessel, and pour on it half a drachm of Nitric Acid mixed with two and a half drachms of Hydrochloric Acid and three drachms of water; digest by a gentle heat, but do not *boil* the acid, or much of the Chlorine will be driven off in the form of gas. At the expiration of a few hours add fresh Aqua Regia in quantity the same as at first, which will probably complete the solution, but if not, repeat the process a third time.

Lastly, neutralize the liquid by adding Carbonate of Soda until all effervescence ceases, and a green precipitate forms; this is *Carbonate of Copper*, which must be allowed several hours to separate thoroughly. The Chloride of Gold is thus freed from Copper and Silver, with which the metallic Gold is alloyed in the standard coin of the realm. The solution so prepared will be *alkaline*, and consequently prone to a reduction of metallic Gold: a slight extra quantity of Hydrochloric acid should therefore be added, sufficient to redden a piece of immersed litmus-paper.

The weight of a half-sovereign is about 61 grains, of which 56 grains are pure Gold. This is equivalent to 86 grains of Chloride of Gold, which will be the quantity contained in the solution.

The following process for preparing Chloride of Gold is more perfect than the last:—Dissolve the Gold coin in Aqua Regia as before; then boil with excess of Hydrochloric Acid, to destroy the Nitric Acid,—dilute largely with distilled water, and add a filtered aqueous solution of common Sulphate of Iron (6 parts to 1 of Gold); collect the precipitated Gold, which is now free from copper; redissolve in Aqua Regia, and evaporate to dryness on a water bath.

Avoid using *Ammonia* to neutralize Chloride of Gold, as it would occasion a deposit of "Fulminating Gold," the properties of which are described in the next page.

Properties of Chloride of Gold.—As sold in commerce it usually contains excess of Hydrochloric Acid, and is then of a bright yellow colour; but when neutral and somewhat concentrated, it is dark red (*Leo ruber* of the alchemists). It gives no precipitate with Carbonate of Soda unless heat be applied; the free Hydrochloric Acid present forms, with the alkali, Chloride of Sodium, which unites with the Chloride of Gold, and produces a double salt, Chloride of Gold and Sodium, soluble in water.

Chloride of Gold is decomposed with precipitation of metallic Gold by Charcoal, Sulphurous Acid, and many of the vegetable acids; also by Protosulphate and Protonitrate of Iron. It tinges the cuticle of an indelible purple tint. It is soluble in Alcohol and in Ether.

GOLD, FULMINATING.

This is a yellowish-brown substance, precipitated on adding Ammonia to a strong solution of Chloride of Gold.

It may be dried carefully at 212°, but *explodes violently* on being heated suddenly to about 290°. Friction also causes it to explode when dry; but the moist powder may be rubbed or handled without danger. It is decomposed by Sulphuretted Hydrogen.

Fulminating Gold is probably an Aurate of Ammonia, containing 2 atoms of Ammonia to 1 atom of Peroxide of Gold.

GOLD, HYPOSULPHITE OF.

Symbol, $AuO\ S_2O_2$. Atomic weight, 253.

Hyposulphite of Gold is produced by the reaction of Chloride of Gold upon Hyposulphite of Soda (see page 133).

The salt sold in commerce as Sel d'or is a double Hyposulphite of Gold and Soda, containing one atom of the former salt to three of the latter, with four atoms of water of crystallization. It is formed by adding one part of Chloride of Gold, in solution, to three parts of Hyposulphite of Soda, and precipitating the resulting salt by Alcohol: the Chloride of Gold must be added to the Hyposulphite of Soda, and not the Soda salt to the Gold (see page 250).

Properties.—Hyposulphite of Gold is unstable and cannot exist in an isolated state, quickly passing into Sulphur, Sulphuric Acid, and metallic Gold. When combined with excess of Hyposulphite of Soda in the form of Sel d'or, it is more permanent.

Sel d'or occurs crystallized in fine needles, which are very soluble in water. The commercial article is often impure, containing little else than Hyposulphite of Soda, with a trace of Gold. It may be analyzed by adding a few drops of strong Nitric Acid (free from Chlorine), diluting with water, and afterwards collecting and igniting the yellow powder, which is metallic Gold.

GRAPE SUGAR.

Symbol, $C_{24}H_{23}O_{23}$. Atomic weight, 396.

This modification of Sugar, often termed *Granular Sugar*, or *Glucose*, exists abundantly in the juice of grapes and in many other varieties of fruit. It forms the saccharine concretion found in honey, raisins, dried figs, etc. It may be produced artificially by the action of fermenting principles and of dilute mineral acids, upon Starch.

Properties.—Grape Sugar crystallizes slowly and with difficulty from a concentrated aqueous solution, in small hemispherical nodules, which are hard, and feel gritty between the teeth. It is much less sweet to the taste than Cane Sugar, and not so soluble in water (1 part dissolves in $1\frac{1}{2}$ of cold water).

Grape Sugar tends to absorb Oxygen, and hence it possesses the property of decomposing the salts of the noble metals, and reducing them by degrees to the metallic state, even without the aid of light. *Cane* Sugar does not possess these properties to an equal extent, and hence it is readily distinguished from the other variety. The product of the action of Grape Sugar upon Nitrate of Silver appears to be a very low form of Oxide of Silver combined with organic matter.

HONEY.

This substance contains two distinct kinds of Sugar, Grape Sugar, and an uncrystallizable substance analogous to, or identical with, the Treacle found associated with common Sugar in the cane-juice. The agreeable taste of Honey probably depends upon the latter, but its reducing power on metallic oxides is due to the former. Pure Grape Sugar can readily be obtained from inspissated Honey, by treating it with Alcohol, which dissolves out the syrup, but leaves the crystalline portion.

Much of the commercial article is adulterated, and, for Photographic use, the Virgin Honey should be obtained direct from the comb.

HYDROCHLORIC ACID.

Symbol, HCl. Atomic weight, 37.

Hydrochloric Acid is a volatile gas, which may be liberated from most of the salts termed Chlorides by the action of Sulphuric Acid. The acid, by its superior affinities, removes the base; thus,—

$$NaCl + HO \, SO_3 = NaO \, SO_3 + HCl.$$

Properties.—Abundantly soluble in water, forming the liquid Hydrochloric or Muriatic Acid of commerce. The most concentrated solution of Hydrochloric Acid has a sp. gr. 1·2, and contains about 40 per cent. of gas; that commonly sold is somewhat weaker, sp. gr. 1·14 = 28 per cent. real acid.

Pure Hydrochloric Acid is colourless, and fumes in the air. The yellow colour of the commercial acid depends upon the presence of traces of Perchloride of Iron, or of organic matter; commercial Muriatic Acid also often contains a portion of free Chlorine and of Sulphuric Acid.

HYDRIODIC ACID.

Symbol, HI.   Atomic weight, 127.

This is a gaseous compound of Hydrogen and Iodine, corresponding in composition to the Hydrochloric Acid. It cannot however, from its instability, be obtained in the same manner, since, on distilling an Iodide with Sulphuric Acid, the Hydriodic Acid first formed is subsequently decomposed into Iodine and Hydrogen. An aqueous solution of Hydriodic Acid is easily prepared by adding Iodine to water containing Sulphuretted Hydrogen gas; a decomposition takes place, and Sulphur is set free: thus, $HS + I = HI + S$.

Properties.—Hydriodic Acid is very soluble in water, yielding a strongly acid liquid. The solution, colourless at first, soon becomes brown from decomposition, and liberation of free Iodine. It may be restored to its original condition by adding solution of Sulphuretted Hydrogen.

HYDROSULPHURIC ACID.

Symbol, HS.   Atomic weight, 17.

This substance, also known as Sulphuretted Hydrogen, is a gaseous compound of Sulphur and Hydrogen, analogous in composition to the Hydrochloric and Hydriodic Acid. It is usually prepared by the action of dilute Sulphuric Acid upon Sulphuret of Iron, as described at page 373; the decomposition being similar to that involved in the preparation of the Hydrogen acids generally:—

$$FeS + HO\ SO_3 = FeO\ SO_3 + HS.$$

Properties.—Cold water absorbs three times its bulk of Hydrosulphuric Acid, and acquires the peculiar putrid odour and poisonous qualities of the gas. The solution is faintly acid to test-paper, and becomes opalescent on keeping, from gradual separation of Sulphur. It is decomposed by Nitric Acid, and also by Chlorine and Iodine.

It precipitates Silver from its solutions in the form of black Sulphuret of Silver; also Copper, Mercury, Lead, etc.; but Iron and other metals of that class are not affected, if the liquid contains free acid. Hydrosulphuric Acid is constantly employed in the chemical laboratory for these and other purposes.

HYDROSULPHATE OF AMMONIA.

Symbol, NH_4S HS. Atomic weight, 51.

The liquid known by this name, and formed on passing Sulphuretted Hydrogen gas into Ammonia, is a double Sulphuret of Hydrogen and Ammonium. In the preparation, the passage of the gas is to be continued until the solution gives no precipitate with Sulphate of Magnesia, and smells strongly of Hydrosulphuric Acid.

Properties.—Colourless at first, but afterwards changes to yellow, from liberation and subsequent solution of Sulphur. Becomes milky on the addition of any acid. Precipitates, in the form of Sulphuret, all the metals which are affected by Sulphuretted Hydrogen, and, in addition, those of the class to which Iron, Zinc, and Manganese belong.

Hydrosulphate of Ammonia is employed in Photography to darken the Negative image, and also in the preparation of Iodide of Ammonium, the separation of Silver from Hyposulphite solutions, etc.

HYPOSULPHITE OF SODA.

Symbol, $NaO\ S_2O_2 + 5\ HO$. Atomic weight, 125.

The chemistry of Hyposulphurous Acid and the Hyposulphite of Soda has been sufficiently described at pages 43, 129, and 137 of the present Work. The crystallized salt includes five atoms of water of crystallization.

HYPOSULPHITE OF GOLD. *See* GOLD, HYPOSULPHITE OF.

HYPOSULPHITE OF SILVER. *See* SILVER, HYPO-
SULPHITE OF.

ICELAND MOSS.

Cetraria Islandica.—A species of Lichen found in Iceland and the mountainous parts of Europe; when boiled
in water, it first swells up, and then yields a substance
which gelatinizes on cooling.

It contains Lichen Starch, a bitter principle soluble in
Alcohol, termed " Cetrarine," and common Starch; traces
of Gallic Acid and Bitartrate of Potash are also present.

IODINE.

Symbol, I. Atomic weight, 126.

Iodine is chiefly prepared at Glasgow, from *kelp*, which
is the fused ash obtained on burning seaweeds. The waters of the ocean contain minute quantities of the Iodides
of Sodium and Magnesium, which are separated and stored
up by the growing tissues of the marine plant.

In the preparation, the mother-liquor of kelp is evaporated to dryness and distilled with Sulphuric Acid; the
Hydriodic Acid first liberated is decomposed by the high
temperature, and fumes of Iodine condense in the form of.
opaque crystals.

Properties.—Iodine has a bluish-black colour and metallic lustre; it stains the skin yellow, and has a pungent
smell, like diluted Chlorine. It is extremely volatile when
moist, boils at 350°, and produces dense violet-coloured
fumes, which condense in brilliant plates. Specific gravity
4·946. Iodine is very sparingly soluble in water, 1 part
requiring 7000 parts for perfect solution; even this minute
quantity however tinges the liquid of a brown colour, Alcohol and Ether dissolve it more abundantly, forming
dark-brown solutions. Iodine also dissolves freely in solutions of the alkaline Iodides, such as the Iodide of Potassium, of Sodium, and of Ammonium.

Chemical Properties.—Iodine belongs to the Chlorine group of elements, characterized by forming acids with Hydrogen, and combining extensively with the metals (see Chlorine). They are however comparatively indifferent to Oxygen, and also to each other. The Iodides of the alkalies and alkaline earths are soluble in water; also those of Iron, Zinc, Cadmium, etc. The Iodides of Lead, Silver, and Mercury are nearly or quite insoluble.

Iodine possesses the property of forming a compound of a deep blue colour with Starch. In using this as a test, it is necessary first to liberate the Iodine (if in combination) by means of Chlorine, or Nitric Acids aturated with Peroxide of Nitrogen. The presence of Alcohol or Ether interferes to a certain extent with the result.

IODIDE OF AMMONIUM.

Symbol, NH_4I. Atomic weight, 144.

The preparation and properties of this salt are described at page 198, to which the reader is referred.

IODIDE OF CADMIUM.

Symbol, CdI. Atomic weight, 182.

See page 199, for the preparation and properties of this salt.

IODIDE OF IRON.

Symbol, FeI. Atomic weight, 154.

Iodide of Iron is prepared by digesting an excess of Iron filings with solution of Iodine in Alcohol. It is very soluble in water and Alcohol, but the solution rapidly absorbs Oxygen and deposits Peroxide of Iron; hence the importance of preserving it in contact with metallic Iron, with which the separated Iodine may recombine. By very careful evaporation, hydrated crystals of Proto-iodide may be obtained, but the composition of the solid salt usually sold under that name cannot be depended on.

The *Periodide* of Iron, corresponding to the *Perchloride*, has not been examined, and it is doubtful if any such compound exists.

IODIDE OF POTASSIUM.

Symbol, KI. Atomic weight, 166.

This salt is usually formed by dissolving Iodine in solution of Potash until it begins to acquire a brown colour; a mixture of Iodide of Potassium and *Iodate of Potash* (KO IO$_5$) is thus formed; but by evaporation and heating to redness, the latter salt parts with its Oxygen, and is converted into Iodide of Potassium.

Properties.—It forms cubic and prismatic crystals, which should be hard, and *very slightly or not at all deliquescent.* Soluble in less than an equal weight of water at 60o; it is also soluble in Alcohol, but not in Ether. The proportion of Iodide of Potassium contained in a saturated alcoholic solution, varies with the strength of the spirit: —with common Spirits of Wine, sp. gr. ·836, it would be about 8 grains to the drachm; with Alcohol rectified from Carbonate of Potash, sp. gr. ·823, 4 or 5 grains; with absolute Alcohol, 1 to 2 grains. The solution of Iodide of Potassium is instantly coloured brown by free Chlorine; also very rapidly by Peroxide of Nitrogen (page 86); ordinary acids however act less quickly, Hydriodic Acid being first formed, and subsequently decomposing spontaneously.

The impurities of commercial Iodide of Potassium, with the means to be adopted for their removal, are fully given at page 197.

IODIDE OF SILVER. *See* SILVER, IODIDE OF.

IODOFORM.

The composition of this substance is analogous to that

of Chloroform, Iodine being substituted for Chlorine. It is obtained on boiling together Iodine, Carbonate of Potash, and Alcohol.

Iodoform occurs in yellow nacrous crystals, which have a saffron-like odour. It is insoluble in water, but soluble in spirit.

IRON, PROTOSULPHATE OF.

Symbol, FeO SO_3+7HO.   Atomic weight, 139.

The properties of this salt, and of the two salifiable Oxides of Iron, are described at page 29. It dissolves in rather more than an equal weight of cold water, or in less of boiling water.

Aqueous solution of Sulphate of Iron absorbs the *Binoxide of Nitrogen*, acquiring a deep olive-brown colour: as this gaseous Binoxide is itself a reducing agent, the liquid so formed has been proposed as a more energetic developer than the Sulphate of Iron alone (?).

IRON, PROTONITRATE OF.

Symbol, FeO NO_5+7 HO.   Atomic weight, 153.

This salt, by careful evaporation *in vacuo* over Sulphuric Acid, forms transparent crystals, of a light green colour, and containing 7 atoms of water, like the Protosulphate. It is exceedingly unstable, and soon becomes red from decomposition, unless preserved from contact with air. The preparation of solution of Protonitrate of Iron for developing Collodion Positives, is given at page 206.

IRON, PERCHLORIDE OF.

Symbol, Fe_2Cl_3.   Atomic weight, 164.

There are two Chlorides of Iron, corresponding in composition to the Protoxide and the Sesquioxide respectively. The Protochloride is very soluble in water, forming a green solution, which precipitates a dirty white Prot-oxide on the addition of an alkali. The Perchloride, on the other hand, is dark brown, and gives a foxy-red precipitate with alkalies.

Properties.—Perchloride of Iron may be obtained in the solid form by heating Iron wire in excess of Chlorine; it condenses in the shape of brilliant and iridescent brown crystals, which are volatile, and dissolve in water, the solution being acid to test-paper. It is also soluble in Alcohol, forming the Tinctura Ferri Sesquichloridi of the Pharmacopœia. Commercial Perchloride of Iron ordinarily contains an excess of Hydrochloric Acid.

LITMUS.

Litmus is a vegetable substance prepared from various *lichens*, which are principally collected on rocks adjoining the sea. The colouring matter is extracted by a peculiar process, and afterwards made up into a paste with chalk, plaster of Paris, etc.

Litmus occurs in commerce in the form of small cubes of a fine violet colour. In using it for the preparation of test-papers, it is digested in hot water, and sheets of porous paper are soaked in the blue liquid so formed. The red papers are prepared at first in the same manner, but afterwards placed in water which has been rendered faintly acid with Sulphuric or Hydrochloric Acid.

MERCURY, BICHLORIDE OF.

Symbol, $HgCl_2$. Atomic weight, 274.

This salt, also called Corrosive Sublimate, and sometimes *Chloride of Mercury* (the atomic weight of Mercury being halved), may be formed by heating Mercury in excess of Chlorine, or more economically, by subliming a mixture of Persulphate of Mercury and Chloride of Sodium.

2 A

Properties.—A very corrosive and poisonous salt, usually sold in semi-transparent, crystalline masses, or in the state of powder. Soluble in 16 parts of cold, and in 3 of hot water; more abundantly so in Alcohol, and also in Ether. The solubility in water may be increased by the addition of free Hydrochloric Acid, or of Chloride of Ammonium.

The Protochloride of Mercury is an insoluble white powder, commonly known under the name of *Calomel.*

METHYLIC ALCOHOL.

This liquid, known also by the names of *wood naphtha* and *pyroxylic spirit*, is one of the volatile products of the destructive distillation of wood. It is very volatile and limpid, with a pungent odour.

By a recent excise regulation, ordinary Spirit mixed with ten per cent. of wood naphtha is sold free of duty, under the name of " Methylated Spirit."

MILK.

The Milk of herbivorous animals contains three principal constituents—Fatty matter, Caseine, and Sugar; in addition to these, small quantities of the Chloride of Potassium, and of Phosphates of Lime and Magnesia, are present.

The fatty matter is contained in small cells, and forms the greater part of the cream which rises to the surface of the milk on standing; hence *skimmed* milk is to be preferred for Photographic use.

The second constituent, *Caseine*, is an organic principle somewhat analogous to Albumen in composition and properties. Its aqueous solution however does not, like Albumen, *coagulate* on boiling, unless *an acid* be present, which probably removes a small portion of alkali with which the Caseine was previously combined. The substance termed " rennet," which is the dried stomach of the calf, possesses the property of coagulating Caseine, but the exact mode of its action is unknown. Sherry wine is also commonly employed to curdle Milk; but brandy and other spirituous liquids, when free from acid and astringent matter, have no effect.

In all these cases a portion of the Caseine usually remains in a soluble form in the *whey;* but when the Milk is coagulated by the addition of acids, the quantity so left is very small, and hence the use of the rennet is to be preferred, since the presence of Caseine facilitates the reduction of the sensitive Silver salts.

Caseine combines with Oxide of Silver in the same manner as Albumen, forming a white coagulum, which becomes *brick-red* on exposure to light.

Sugar of Milk, the third principal constituent, differs from both cane and grape sugar; it may be obtained by evaporating *whey* until crystallization begins to take place. It is hard and gritty, and only slightly sweet; slowly soluble, without forming a syrup, in about two and a half parts of boiling, and six of cold water. It does not ferment and form Alcohol on the addition of yeast, like grape sugar, but by the action of *decomposing animal matter* is converted into Lactic Acid.

When skimmed Milk is exposed to the air for some hours, it gradually becomes *sour*, from Lactic Acid formed in this way; and if then heated to ebullition, the Caseine coagulates very perfectly.

NITRIC ACID.

Symbol, NO_5. Atomic weight, 54.

Nitric Acid, or *Aqua-fortis*, is prepared by adding Sulphuric Acid to Nitrate of Potash, and distilling the mixture in a retort. Sulphate of Potash and free Nitric Acid are formed, the latter of which, being volatile, distils over in combination with one atom of water previously united with the Sulphuric Acid.

Properties.—Anhydrous Nitric Acid is a solid substance, white and crystalline, but it cannot be prepared except by an expensive and complicated process.

The concentrated *liquid* Nitric Acid contains 1 atom of water, and has a sp. gr. of about 1·5; if perfectly pure, it is colourless, but usually it has a slight yellow tint, from partial decomposition into Peroxide of Nitrogen: it fumes strongly in the air.

The strength of commercial Nitric Acid is subject to much variation. An acid of sp. gr. 1·42, containing about 4 atoms of water, is commonly met with. If the specific gravity is much lower than this (less than 1·36), it will scarcely be adapted for the preparation of Pyroxyline. The yellow *Nitrous Acid,* so called, is a strong Nitric Acid partially saturated with the brown vapours of Peroxide of Nitrogen; it has a high specific gravity, but this is somewhat deceptive, being caused in part by the presence of the Peroxide. On mixing with Sulphuric Acid, the colour disappears, a compound being formed which has been termed a *Sulphate of Nitrous Acid.*

In the Appendix a Table is given which exhibits the quantity of real anhydrous Nitric Acid contained in samples of different densities.

Chemical Properties.—Nitric Acid is a powerful oxidizing agent (see page 13); it dissolves all the common metals, with the exception of Gold and Platinum. Animal substances, such as the cuticle, nails, etc., are tinged of a permanent yellow colour, and deeply corroded by a prolonged application. Nitric Acid forms a numerous class of salts, *all of which are soluble in water.* Hence its presence cannot be determined by any precipitating reagent, in the same manner as that of Hydrochloric and Sulphuric Acid.

Impurities of Commercial Nitric Acid.—These are principally *Chlorine* and *Sulphuric Acid;* also Peroxide of Nitrogen, which tinges the acid yellow, as already described. Chlorine is detected by diluting the acid with an equal bulk of distilled water, and adding a few drops of Nitrate of Silver,—*a milkiness*, which is Chloride of Silver in suspension, indicates the presence of Chlorine. In testing for Sulphuric Acid, dilute the Nitric Acid as before, and drop in *a single drop* of solution of Chloride of Barium; if Sulphuric Acid be present, an insoluble precipitate of Sulphate of Baryta will be formed.

NITROUS ACID. *See* SILVER, NITRITE OF.

NITRATE OF POTASH.

Symbol, $KO\ NO_5$. Atomic weight, 102.

This salt, also termed *Nitre*, or *Saltpetre*, is an abundant natural product, found effloresced upon the soil in certain parts of the East Indies. It is also produced artificially in what are called Nitre-beds.

The properties of Nitrate of Potash are described as far as necessary at page 190.

NITRATE OF BARYTA.

Symbol, $BaO\ NO_5$. Atomic weight, 131.

Nitrate of Baryta forms octahedral crystals, which are anhydrous. It is considerably less soluble than the Chloride of Barium, requiring 12 parts of cold and 4 of boiling water for solution. It may be substituted for the Nitrate of Lead in the preparation of Protonitrate of Iron.

NITRATE OF LEAD.

Symbol, $PbO\ NO_5$. Atomic weight, 166.

Nitrate of Lead is obtained by dissolving the metal, or the Oxide of Lead, in *excess* of Nitric Acid, diluted with 2 parts of water. It crystallizes on evaporation in white anhydrous tetrahedra and octahedra, which are hard, and decrepitate on being heated; they are soluble in 8 parts of water at 60°.

Nitrate of Lead forms with Sulphuric Acid, or soluble Sulphates, a white precipitate, which is the insoluble Sulphate of Lead. The *Iodide* of Lead is also very sparingly soluble in water.

NITRATE OF SILVER. *See* SILVER, NITRATE OF.

NITRO-GLUCOSE.

When 3 fluid ounces of cold Nitro-Sulphuric Acid, consisting of 2 ounces of Oil of Vitriol and 1 ounce of highly concentrated Nitric Acid, are mixed with 1 ounce of finely powdered Cane Sugar, there is formed at first a thin, transparent, pasty mass. If it is stirred with a glass rod for a few minutes without interruption, the paste coagulates as it were, and separates from the liquid as a thick tenacious mass, aggregating into lumps, which can easily be removed from the acid mixture.

This substance has a very acid and intensely bitter taste. Kneaded in warm water until the latter no longer reddens litmus-paper, it acquires a silver colour and a beautiful silky lustre. It may be used in Photography to confer intensity upon newly mixed Collodion; but is inferior to Glycyrrhizine employed for the same purpose.

NITRO-HYDROCHLORIC ACID.

Symbol, $NO_4 + Cl$.

This liquid is the Aqua-regia of the old alchemists. It is produced by mixing Nitric and Hydrochloric Acids: the Oxygen contained in the former combines with the Hydrogen of the latter, forming water and liberating Chlorine, thus :—

$$NO_5 + HCl = NO_4 + HO + Cl.$$

The presence of free Chlorine confers on the mixture the power of dissolving Gold and Platinum, which neither of the two acids possesses separately. In preparing Aqua-regia it is usual to mix one part, by measure, of Nitric Acid with four of Hydrochloric Acid, and to dilute with an equal bulk of water. The application of a gentle heat assists the solution of the metal; but if the temperature rises to the boiling point, a violent effervescence and escape of Chlorine takes place.

NITRO-SULPHURIC ACID.

For the chemistry of this acid liquid, see page 77.

OXYGEN.

Symbol, O. Atomic weight, 8.

Oxygen gas may be obtained by heating Nitrate of Potash to redness, but in this case it is contaminated with a portion of Nitrogen. The salt termed Chlorate of Potash (the composition of which is closely analogous to that of the Nitrate, Chlorine being substituted for Nitrogen) yields abundance of pure Oxygen gas on the application of heat, leaving behind Chloride of Potassium.

Chemical Properties.—Oxygen combines eagerly with many of the chemical elements, forming Oxides. This chemical affinity however is not well seen when the elementary body is exposed to the action of *Oxygen in the gaseous form*. It is the *nascent* Oxygen which acts most powerfully as an oxidizer. By nascent Oxygen is meant Oxygen on the point of separation from other elementary atoms with which it was previously associated; it may then be considered to be in the liquid form, and hence it comes more perfectly into contact with the particles of the body to be oxidized.

Illustrations of the superior chemical energy of nascent Oxygen are numerous, but none perhaps are more striking than the mild and gradual oxidizing influence exerted by atmospheric air, as compared with the violent action of

Nitric Acid and bodies of that class which contain Oxygen loosely combined.

OXYMEL.

This syrup of Honey and Vinegar is prepared as follows.   Take of

Honey . . . . 1 pound.
Acid, Acetic, fortiss. (Beaufoy's Acid) 11 drachms.
Water . . . . 13 drachms.

Stand the pot containing the Honey in boiling water until a scum rises to the surface, which is to be removed two or three times.   Then add the Acetic Acid and water, and skim once more if required.   Allow to cool, and it will be fit for use.

POTASH.

Symbol, KO + HO.   Atomic weight, 57.

Potash is obtained by separating the Carbonic Acid from Carbonate of Potash by means of Caustic Lime. Lime is a more feeble base than Potash, but the Carbonate of Lime, being *insoluble* in water, is at once formed on adding Milk of Lime to a solution of Carbonate of Potash (see page 314).

Properties.—Usually met with in the form of solid lumps, or in cylindrical sticks, which are formed by melting the Potash and running it into a mould.   It always contains one atom of water, which cannot be driven off by the application of heat.

Potash is soluble almost to any extent in water, much heat being evolved.   The solution is powerfully alkaline (p. 308), and acts rapidly upon the skin; it dissolves fatty and resinous bodies, converting them into soaps. Solution of Potash absorbs Carbonic Acid quickly from the air, and should therefore be preserved in stoppered bottles; the glass stoppers must be wiped occasionally, in order to prevent them from becoming immovably fixed by the solvent action of the Potash upon the Silica of the glass.

The Liquor Potassæ of the London Pharmacopœia has a sp. gr. of 1·063, and contains about 5 per cent. of real Potash. It is usually contaminated with *Carbonate* of Potash, which causes it to effervesce on the addition of acids; also, to a less extent, with Sulphate of Potash, Chloride of Potassium, Silica, etc.

POTASH, CARBONATE OF.

Symbol, $KO\ CO_2$. Atomic weight, 70.

The impure Carbonate of Potash, termed *Pearlash*, is obtained from the ashes of wood and vegetable matter, in the same manner as Carbonate of Soda is prepared from the ashes of seaweeds. Salts of Potash and of Soda appear essential to vegetation, and are absorbed and approximated by the living tissues of the plant. They exist in the vegetable structure, combined with organic acids in the form of salts, like the Oxalate, Tartrate, etc., which, when burned are converted into Carbonates.

Properties.—The Pearlash of commerce contains large and variable quantities of Chloride of Potassium, Sulphate of Potash, etc. A purer Carbonate is sold, which is free from Sulphates, and with only a trace of Chlorides. Carbonate of Potash is a strongly alkaline salt, deliquescent, and soluble in twice its weight of cold water; insoluble in Alcohol, and employed to deprive it of water (see page 196).

PYROGALLIC ACID.

Symbol, $C_8H_4O_4$ (Stenhouse). Atomic weight, 84.

The chemistry of Pyrogallic Acid has been described at page 28.

SEL D'OR. *See* GOLD, HYPOSULPHITE OF.

SILVER.

Symbol, Ag. Atomic weight, 108.

This metal, the *Luna* or *Diana* of the alchemists, is found native in Peru and Mexico; it occurs also in the form of Sulphuret of Silver.

When pure it has a sp. gr. of 10·5, and is very malleable and ductile; melts at a bright red heat. Silver does not oxidize in the air, but when exposed to an impure atmosphere containing traces of Sulphuretted Hydrogen, it is slowly tarnished from formation of Sulphuret of Silver. It dissolves in Sulphuric Acid, but the best solvent is Nitric Acid.

The standard coin of the realm is an alloy of Silver and Copper, containing about one-eleventh of the latter metal.

To prepare pure Nitrate of Silver from it, dissolve in Nitric Acid and evaporate until crystals are obtained. Then wash the crystals with a little dilute Nitric Acid, redissolve them in water, and crystallize by evaporation a second time. Lastly, fuse the product at a moderate heat, in order to expel the last traces of Nitric and Nitrous Acids.

SILVER, AMMONIO-NITRATE OF.

Crystallized Nitrate of Silver absorbs Ammoniacal gas rapidly, with production of heat sufficient to fuse the resulting compound, which is white, and consists of 100 parts of the Nitrate + 29·5 of Ammonia. The compound however which Photographers employ under the name of Ammonio-Nitrate of Silver may be viewed more simply as a solution of the Oxide of Silver in Ammonia, without reference to the Nitrate of Ammonia necessarily produced in the reaction.

Very strong Ammonia, in acting upon Oxide of Silver,

converts it into a black powder, termed *Fulminating Silver*, which possesses the most dangerous explosive properties. Its composition is uncertain. In preparing Ammonio-Nitrate of Silver by the common process, the Oxide first precipitated occasionally leaves a little black powder behind, on re-solution; this does not appear however, according to the observations of the Author, to be Fulminating Silver.

In sensitizing salted paper by the Ammonio-Nitrate of Silver, *free Ammonia* is necessarily formed. Thus—

Chloride of Ammonium + Oxide of Silver in Ammonia
= Chloride of Silver   + Ammonia + Water.

SILVER, OXIDE OF.

Symbol, AgO. Atomic weight, 116.

This compound has already been described in Part I., page 17.

SILVER, CHLORIDE OF.

Symbol, AgCl. Atomic weight, 144.

The preparation and properties of Chloride of Silver are given in Part I. page 14.

SILVER, BROMIDE OF.

Symbol, AgBr. Atomic weight, 186.

See Part I. page 17.

SILVER, CITRATE OF. *See* CITRIC ACID.

SILVER, IODIDE OF.

Symbol, AgI. Atomic weight, 234.

See Part I. page 16.

SILVER, FLUORIDE OF.

Symbol, AgF. Atomic weight, 127.

This compound differs from those last described in being soluble in water. The dry salt fuses on being heated, and is reduced by a higher temperature, or by exposure to light.

SILVER, SULPHURET OF.

Symbol, AgS. Atomic weight, 124.

This compound is formed by the action of Sulphur upon metallic Silver, or of Sulphuretted Hydrogen or Hydrosulphate of Ammonia upon the Silver salts; the decomposition of Hyposulphite of Silver also furnishes the black Sulphuret.

Sulphuret of Silver is insoluble in water, and nearly so in those substances which dissolve the Chloride, Bromide, and Iodide, such as Ammonia, Hyposulphites, Cyanides, etc.; but it dissolves in Nitric Acid, being converted into soluble Sulphate and Nitrate of Silver. (For a further account of the properties of the Sulphuret of Silver, see page 146.)

SILVER, NITRATE OF.

Symbol, AgO NO$_5$. Atomic weight, 170.

The preparation and properties of this salt have been explained at pages 12 and 362.

SILVER, NITRITE OF.

Symbol, AgO NO$_3$. Atomic weight, 154.

Nitrite of Silver is a compound of Nitrous Acid, or NO$_3$, with Oxide of Silver. It is formed by heating Nitrate of Silver, so as to drive off a portion of its Oxygen, or more conveniently, by mixing Nitrate of Silver and Nitrite of Potash in equal parts, fusing strongly, and dissolving in a small quantity of boiling water : on cooling, the Nitrite crystallizes out, and may be purified by pressing in blotting-paper. Mr. Hadow describes an economical method of preparing Nitrite of Silver in quantity, viz. by heating 1 part of Starch in 8 of Nitric Acid of 1·25 specific gravity, and conducting the evolved gases into a solution of pure Carbonate of Soda until effervescence has ceased. The Nitrite of Soda thus formed is afterwards added to Nitrate of Silver in the usual way.

Properties.—Nitrite of Silver is soluble in 120 parts of cold water ; easily soluble in boiling water, and crystallizes, on cooling, in long slender needles. It has a certain degree of affinity for Oxygen, and tends to pass into the condition of Nitrate of Silver ; but it is probable that its Photographic properties depend more upon a decomposition of the salt and liberation of Nitrous Acid.

Properties of Nitrous Acid.—This substance possesses very feeble acid properties, its salts being decomposed even by Acetic Acid. It is an unstable body, and splits up, in contact with water, into Binoxide of Nitrogen and Nitric Acid. The Peroxide of Nitrogen, NO_4, is also decomposed by water, and yields the same products.

SILVER, ACETATE OF.

Symbol, AgO ($C_4H_3O_3$). Atomic weight, 167.

This is a difficultly soluble salt, deposited in lamellar crystals when an Acetate is added to a strong solution of Nitrate of Silver. If *Acetic Acid* be used in place of an Acetate, the Acetate of Silver does not fall so readily, since the Nitric Acid which would then be liberated impedes the decomposition. Its properties have been sufficiently described at page 89.

SILVER, HYPOSULPHITE OF.

Symbol, $AgO\ S_2O_2$. Atomic weight, 164.

This salt is fully described in Part I. page 129. For the properties of the soluble double salt of Hyposulphite of Silver and Hyposulphite of Soda, see page 43.

SUGAR OF MILK. *See* MILK.

SULPHURETTED HYDROGEN. *See* HYDROSULPHURIC ACID.

SULPHURIC ACID.

Symbol, SO_3. Atomic weight, 40.

Sulphuric Acid may be formed by oxidizing Sulphur with boiling Nitric Acid; but this plan would be too expensive to be adopted on a large scale. The commercial process for the manufacture of Sulphuric Acid is exceedingly ingenious and beautiful, but it involves reactions which are too complicated to admit of a superficial explanation. The Sulphur is first burnt into gaseous Sulphurous Acid (SO_2), and then by the agency of Binoxide of Nitrogen gas, an additional atom of Oxygen is imparted from the atmosphere, so as to convert the SO_2 into SO_3, or Sulphuric Acid.

Properties.—Anhydrous Sulphuric Acid is a white crystalline solid. The strongest liquid acid always contains one atom of water, which is closely associated with it, and cannot be driven off by the application of heat.

This *mono-hydrated* Sulphuric Acid, represented by the formula $HO\ SO_3$, is a dense fluid, having a specific gravity of about 1·845; boils at 620°, and distils without decomposition. It is not volatile at common temperatures, and therefore does not *fume* in the same manner as Nitric or Hydrochloric Acid. The concentrated acid may be cooled down even to zero without solidifying; but a weaker compound, containing twice the quantity of water, and termed *glacial* Sulphuric Acid, crystallizes at 40° Fahr. Sulphuric Acid is intensely acid and caustic, but it does not destroy the skin or dissolve metals so readily as Nitric Acid. It has an energetic attraction for water, and when the two are mixed, condensation ensues, and much heat is evolved; four parts of acid and one of water produce a temperature equal to that of boiling water. Mixed with aqueous Nitric Acid, it forms the compound known as Nitro-Sulphuric Acid.

Sulphuric Acid possesses intense chemical powers, and displaces the greater number of ordinary acids from their salts. It *chars* organic substances, by removing the elements of water, and converts Alcohol into Ether in a similar manner. The *strength* of a given sample of Sulphuric Acid may be calculated, nearly, from its specific gravity, and a Table is given by Dr. Ure for that purpose. (See Appendix.)

Impurities of Commercial Sulphuric Acid.—The liquid acid sold as *Oil of Vitriol* is tolerably constant in composition, and seems to be as well adapted for Photographic use as the *pure* Sulphuric Acid, which is far more expensive. The specific gravity should be about 1·836 at 60°. If a drop, evaporated upon Platinum foil, gives a fixed residue, probably Bisulphate of Potash is present. A milkiness, on dilution, indicates Sulphate of Lead (see page 186).

Test for Sulphuric Acid.—If the presence of Sulphuric Acid, or a soluble Sulphate, be suspected in any liquid, it is tested for by adding a few drops of dilute solution of Chloride of Barium, or Nitrate of Baryta. A white precipitate, *insoluble in Nitric Acid*, indicates Sulphuric Acid. If the liquid to be tested is very acid, from Nitric or Hydrochloric Acid, it must be largely diluted before testing, or a crystalline precipitate will form, caused by the sparing solubility of the Chloride of Barium itself in acid solutions.

SULPHUROUS ACID.

Symbol, SO_2. Atomic weight, 32.

This is a gaseous compound, formed by burning Sulphur in atmospheric air or Oxygen gas : also by heating Oil of Vitriol in contact with metallic Copper, or with Charcoal.

When an acid of any kind is added to Hyposulphite of Soda, Sulphurous Acid is formed as a product of the decomposition of Hyposulphurous Acid, but it afterwards disappears from the liquid by a secondary reaction, resulting in the production of Trithionate and Tetrathionate of Soda.

Properties.—Sulphurous Acid possesses a peculiar and suffocating odour, familiar to all in the fumes of burning Sulphur. It is a feeble acid, and escapes with effervescence, like Carbonic Acid, when its salts are treated with Oil of Vitriol. It is soluble in water.

TETRATHIONIC ACID.

Symbol, S_4O_5. Atomic weight, 104.

The chemistry of the Polythionic Acids and their salts will be found described in the First Part of this Work, page 157.

WATER.

Symbol, HO. Atomic weight, 9.

Water is an Oxide of Hydrogen, containing single atoms of each of the gases.

Distilled water is water which has been vaporized and again condensed ; by this means it is freed from earthy and saline impurities, which, not being volatile, are left in the body of the retort. *Pure* distilled water leaves no residue on evaporation, and should remain perfectly clear on the addition of Nitrate of Silver, *even when exposed to the light;* it should also be neutral to test-paper.

The condensed water of steam-boilers sold as distilled water is apt to be contaminated with oily and empyreumatic matter, which discolours Nitrate of Silver, and is therefore injurious.

Rain-water, having undergone a natural process of distillation, is free from inorganic salts, but it usually contains a minute portion of *Ammonia*, which gives it an alkaline reaction to test-paper. It is very good for Photographic purposes if collected in clean vessels, but when taken from a common rain-water tank should always be examined, and if much organic matter be present, tingeing it of a brown colour and imparting an unpleasant smell, it must be rejected.

Spring or *River* water, commonly known as "hard water," usually contains Sulphate of Lime, and Carbonate of Lime dissolved in Carbonic Acid; also Chloride of Sodium in greater or less quantity. On boiling the water, the Carbonic Acid gas is evolved, and the greater part of the Carbonate of Lime (if any is present) deposits, forming an earthy incrustation on the boiler.

In testing water for Sulphates and Chlorides, acidify a portion with a few drops of *pure* Nitric Acid, free from Chlorine (if this is not at hand, use pure Acetic Acid); then divide it into two parts, and add to the first a *dilute* solution of Chloride of Barium, and to the second, Nitrate of Silver,—a milkiness indicates the presence of Sulphates in the first case or of Chlorides in the second. The *Photographic Nitrate Bath* cannot be used as a test, since the Iodide of Silver it contains is precipitated on dilution, giving a milkiness which might be mistaken for Chloride of Silver.

Common hard water can often be used for making a Nitrate Bath when nothing better is at hand. The Chlorides it contains are precipitated by the Nitrate of Silver, leaving soluble *Nitrates* in solution, which are not injurious. The Carbonate of Lime, if any is present, neutralizes free Nitric Acid, rendering the Bath alkaline in the same

2 B

manner as Carbonate of Soda. (See page 89.) Sulphate of Lime, usually present in well water, is said to exercise a retarding action upon the sensitive Silver Salts, but on this point the writer is unable to give certain information.

Hard water is not often sufficiently pure for the developing fluids. The Chloride of Sodium it contains decomposes the Nitrate of Silver upon the film, and the image cannot be brought out perfectly. The *New River water*, however, supplied to many parts of London, is almost free from Chlorides, and answers very well. In other cases a few drops of Nitrate of Silver solution may be added, to separate the Chlorine, taking care not to use a large excess.

APPENDIX.

————◆————

QUANTITATIVE TESTING OF SOLUTIONS OF NITRATE OF
SILVER.

THE amount of Nitrate of Silver contained in solutions of that salt
may be estimated with sufficient delicacy for ordinary Photographic
operations by the following simple process.

Take the *pure* crystallized Chloride of Sodium, and either dry it
strongly or fuse it at a moderate heat, in order to drive off any water
which may be retained between the interstices of the crystals ; then
dissolve in distilled water, in the proportion of 8½ grains to 6 fluid
ounces.

In this way, a standard solution of salt is formed, each drachm of
which (containing slightly more than one-sixth of a grain of salt) will
precipitate exactly half a grain of Nitrate of Silver.

In order to use it, measure out accurately one drachm of the Bath in
a minim measure and place it in a two-ounce stoppered phial, taking
care to rinse out the measure with a drachm of distilled water, which
is to be added to the former ; then pour in the salt solution, in the
proportion of a drachm for every 4 grains of Nitrate *known to be
present* in an ounce of the Bath which is to be tested ; shake the contents of the bottle briskly, until the white curds have perfectly separated, and the supernatant liquid is clear and colourless ; then add
fresh portions of the standard solution, by 30 minims at a time, with
constant shaking. When the last addition causes no *milkiness*, read
off the total number of drachms employed (the last half-drachm being

subtracted), and multiply that number by 4 for the weight in grains of the Nitrate of Silver present in an ounce of the Bath.

In this manner the strength of the Bath is indicated within two grains to the ounce, or even to a single grain if the last additions of standard salt-solution be made in portions of 15, instead of 30 minims.

Supposing the Bath to be tested is thought to contain about 35 grains of Nitrate to the ounce, it will be convenient to begin by adding to the measured drachm, 7 *drachms* of the standard solution; afterwards, as the milkiness and precipitation become less marked, the process must be carried on more cautiously, and the bottle shaken violently for several minutes, in order to obtain a clear solution. A few drops of Nitric Acid added to the Nitrate of Silver facilitate the deposition of the Chloride; but care must be taken that the sample of Nitric Acid employed is pure and free from Chlorine, the presence of which would cause an error.

RECOVERY OF SILVER FROM WASTE SOLUTIONS,—FROM THE BLACK DEPOSIT OF HYPO-BATHS, ETC.

The manner of separating metallic Silver from waste solutions varies according to the presence or absence of alkaline Hyposulphites and Cyanides.

a. *Separation of metallic Silver from old Nitrate Baths.*—The Silver contained in solutions of the Nitrate, Acetate, etc. may easily be precipitated by suspending a strip of sheet Copper in the liquid; the action is completed in two or three days, the whole of the Nitric Acid and Oxygen passing to the Copper, and forming a blue solution of the Nitrate of Copper. The metallic Silver however, separated in this manner, always contains a portion of Copper, and gives a blue solution when dissolved in Nitric Acid.

A better process is to commence by precipitating the Silver entirely in the form of *Chloride of Silver*, by adding common Salt until no further milkiness can be produced. If the liquid is well stirred, the Chloride of Silver sinks to the bottom, and may be washed by repeatedly filling the vessel with common water, and pouring off the upper clear portion when the clots have again settled down. The Chloride of Silver thus formed may afterwards be reduced to metallic Silver by a process which will presently be described (p. 374).

b. *Separation of Silver from solutions containing alkaline Hypo-sulphites, Cyanides, or Iodides.*—In this case the Silver cannot be precipitated by adding Chloride of Sodium, since the Chloride of Silver is *soluble* in such liquids. It is necessary therefore to use the Sulphuretted Hydrogen, or the Hydrosulphate of Ammonia, and to separate the Silver in the form of *Sulphuret.*

Sulphuretted Hydrogen gas is readily prepared, by fitting a cork and flexible tubing to the neck of a pint bottle, and having introduced *Sulphuret of Iron* (sold by operative chemists for the purpose), about as much as will stand in the palm of the hand, pouring upon it $1\frac{1}{2}$ fluid ounce of Oil of Vitriol diluted with 10 ounces of water. The gas is generated gradually without the application of heat, and must be allowed to bubble up through the liquid from which the Silver is to be separated. The smell of Sulphuretted Hydrogen being offensive, and highly poisonous if inhaled in a concentrated form, the operation must be carried on in the open air, or in a place where the fumes may escape without doing injury.

When the liquid begins to acquire a strong and persistent odour of Sulphuretted Hydrogen, the precipitation of Sulphuret is completed. The black mass must then be collected upon a filter, and washed by pouring water over it, until the liquid which runs through gives little or no precipitate with a drop of Nitrate of Silver.

The Silver may also be separated in the form of Sulphuret from old Hypo-Baths, by adding Oil of Vitriol in quantity sufficient to decompose the Hyposulphite of Soda; and burning off the free Sulphur from the brown deposit.

Conversion of Sulphuret of Silver into metallic Silver.—The black Sulphuret of Silver may be reduced to the state of metal by roasting and subsequent fusion with Carbonate of Soda; but it is more convenient, in operating on a small scale, to proceed in the following manner :—first convert the Sulphuret into Nitrate of Silver, by boiling with Nitric Acid diluted with two parts of water; when all evolution of red fumes has ceased, the liquid may be diluted, allowed to cool, and filtered from the insoluble portion, which consists principally of Sulphur, but also contains a mixture of Chloride and Sulphuret of Silver, unless the Nitric Acid employed was free from Chlorine; this precipitate may be heated, in order to volatilize the Sulphur, and then digested with Hyposulphite of Soda, or added to the Hypo-Bath.

The solution of Nitrate of Silver obtained by dissolving Sulphuret of Silver, is always strongly acid with Nitric Acid, and also contains *Sulphate* of Silver. It may be crystallized by evaporation ; but unless the quantity of material operated on is large, it will be better to precipitate the Silver in the form of Chloride, by adding common Salt, as already recommended.

REDUCTION OF CHLORIDE OF SILVER TO THE METALLIC STATE.

The Chloride of Silver is first to be carefully washed, by filling up the vessel which contains it, many times with water, and pouring off the liquid, or drawing it off close with a siphon. It may then be dried at a gentle heat, and fused with twice its weight of dry Carbonate of Potash, or better still, with a mixture of the Carbonates of Potash and Soda.

The process for reducing Chloride of Silver in the moist way, by metallic Zinc and Sulphuric Acid, is more economical and less troublesome than that just given ; it is conducted as follows :—The Chloride, after having been well washed as before, is placed in a large flat dish, and a bar of metallic Zinc laid in contact with it. A small quantity of Oil of Vitriol, diluted with four parts of water, is then added, until a slight effervescence of Hydrogen gas is seen to take place. The vessel is set aside for two or three days, and is not to be disturbed, either by stirring or by moving the bar. The reduction begins with the Chloride immediately in contact with the Zinc, and radiates in all directions. When the whole mass has become of a grey colour, the bar is to be carefully removed and the adhering Silver washed off with a stream of water ; the Zinc usually presents a honeycombed appearance, with irregularities upon the surface, which however are not metallic Silver ;—they consist only of Zinc or of Oxide of Zinc.

In order to ensure the purity of the Silver, a fresh addition of Sulphuric Acid must be made, after the Zinc bar has been removed, and the digestion continued for several hours, in order to dissolve any fragments of metallic Zinc which may have been inadvertently detached. The grey powder must be repeatedly washed, first with Sulphuric Acid and water (this is necessary to dissolve a portion of an insoluble Salt of Zinc, probably an oxychloride) and then with water alone, until the liquid runs away *neutral*, and gives no precipitate with Carbonate of Soda; it may then be fused into a button, to burn off organic matter if present, and subsequently converted into Nitrate of Silver by boiling with Nitric Acid diluted with two parts of water.

In reducing Chloride of Silver precipitated from old Nitrate Baths *containing Iodide of Silver,* the grey metallic powder is sometimes contaminated with unreduced Iodide of Silver, which dissolves in the solution of Nitrate of Silver formed on treating the mass with Nitric Acid. To avoid this, wash the purified Silver with solution of Hyposulphite of Soda, and then again with water.

MODE OF TAKING THE SPECIFIC GRAVITY OF LIQUIDS.

Instruments are sold, termed "Hydrometers," which indicate specific gravity by the extent to which a glass bulb containing air, and properly balanced, rises or sinks, in the liquid; but a more exact process, and one equally simple, is by the use of the specific gravity bottle.

These bottles are made to contain exactly 1000 grains of distilled water, and with each is sold *a brass weight,* which counterbalances it when filled with pure water.

In taking the specific gravity of a liquid, fill the bottle quite full and insert the stopper, which being pierced through by a fine capillary tube allows the excess to escape. Then, having wiped the bottle quite dry, place it in the scale-pan, and ascertain the number of grains required to produce equilibrium; this number added to, or subtracted from, *unity* (the assumed specific gravity of water), will give the density of the liquid.

Thus, to take examples, supposing the bottle filled with *rectified Ether* to require 250 grains to enable it to counterbalance the brass weight,—then 1· *minus* ·250, or ·750, is the specific gravity; but in the case of *Oil of Vitriol* the bottle, when full, will be *heavier* than the counterpoise by perhaps 836 grains; therefore 1· *plus* ·836, *id est* 1·836, is the density of the sample examined.

Sometimes the bottle is made to hold only 500 grains of distilled water, in place of 1000; in this case the number of grains to be added or subtracted must be multiplied by 2.

In taking specific gravities, observe that the temperature be within a few degrees of 60° Fahrenheit (if higher or lower, immerse the bottle in warm or cold water); and wash out the bottle thoroughly with water each time after use.

ON FILTRATION AND WASHING PRECIPITATES.

In preparing filters, cut the paper into squares of a sufficient size, and fold each square neatly upon itself, first into a half-square, and then again, at right angles, into a quarter-square;—round off the corners with a pair of scissors, and open out the filter into a conical form, when it will be found to drop exactly into the funnel, and to be uniformly supported throughout.

Before pouring in the liquid, always moisten the filter with distilled water, in order to expand the fibres; if this precaution be neglected, the pores are apt to become choked in filtering liquids which contain finely divided matter in suspension. The solution to be filtered may be poured gently down a glass rod, held in the left hand (*a silver spoon* may be used, in case of necessity, for Nitrate Baths, and all liquids not containing Nitric or Hydrochloric Acid), and directed against the side of the funnel, near to the upper part. If it does not immediately run clear, it will usually do so on returning it into the filter and allowing it to pass through a second time.

Mode of Washing Precipitates.—Collect the precipitate upon a filter and drain off as much of the mother-liquor as possible; then pour in distilled water by small portions at a time, allowing each to percolate through the deposit before adding a fresh quantity. When the water passes through perfectly pure, the washing is complete; in testing it, a single drop may be laid upon a strip of glass and allowed to evaporate spontaneously in a warm place, or the proper chemical reagents may be applied, and the washing continued until no impurity can be detected. Thus, for example, in washing the Sulphuret of Silver precipitated from a Hypo-Bath by means of Hydrosulphate of Ammonia, the process will be completed when the water which runs through causes no deposit with a drop of Nitrate of Silver solution.

ON THE USE OF TEST-PAPERS.

The nature of the colouring matter which is employed in the preparation of litmus-paper has already been described at page 353.

In testing for the alkalies and basic oxides generally, the blue litmus-paper which has been reddened by an acid may be used, or, in place of it, the *turmeric*-paper. Turmeric is a yellow vegetable substance which possesses the property of becoming brown when treated with an alkali; it is however less sensitive than the reddened litmus, and is scarcely affected by the weaker bases, such as Oxide of Silver.

In using test-papers, observe the following precautions :—they should be kept in a dark place, and protected from the action of the air, or they soon become purple from Carbonic Acid, always present in the atmosphere in small quantity. By immersion in water containing about one drop of Liquor Potassæ or Ammoniæ, or a grain of Carbonate of Soda to four ounces, the blue colour is restored. As the quantities which are tested for in Photography are often infinitesimally small, it is essential that the litmus-paper should be in good condition; and test-papers prepared with *porous* paper will be found to show the colour better than those upon glazed or strongly-sized paper. The mode of employing the paper is as follows:—Place a small strip in the liquid to be examined : if it becomes at once *bright red*, a strong acid is present; but if it changes *slowly* to a *wine-red* tint, a weak acid, such as Acetic or Carbonic, is indicated. In the case of the Photographic Nitrate Bath faintly acidified with Acetic Acid, a purple colour only may be expected, and a decided red colour would suggest the presence of Nitric Acid. In the Hypo fixing and toning Bath which has acquired acidity, the litmus-paper will perhaps redden in about three or four minutes.

Blue litmus-papers may be changed to the red papers used for alkalies by soaking in water acidified with Sulphuric Acid, one drop to half a pint; or by holding for an instant near the mouth of a bottle containing Glacial Acetic Acid. In examining a Nitrate Bath for alkalinity by means of the reddened litmus-paper, at least five or ten minutes should be allowed for the action, since the change of colour from red to blue takes place very slowly.

REMOVAL OF SILVER STAINS FROM THE HANDS, LINEN, ETC.

The black stains upon the hands caused by Nitrate of Silver, may readily be removed by moistening them and rubbing with a lump of Cyanide of Potassium. As this salt however is highly poisonous, many may prefer the following plan :—Wet the spot with a saturated solution of Iodide of Potassium, and afterwards with Nitric Acid (the strong Nitric Acid acts upon the skin and turns it yellow, it must therefore be diluted with two parts of water before use) ; then wash with solution of Hyposulphite of Soda.

Stains upon white linen may be easily removed by brushing them with a solution of Iodine in Iodide of Potassium, and afterwards washing with water and soaking in Hyposulphite of Soda, or Cyanide of Potassium, until the yellow Iodide of Silver is dissolved out; the Bichloride of Mercury (neutral solution) also answers well in many cases, changing the dark spot to white (p. 151).

A TABLE SHOWING THE QUANTITY OF ANHYDROUS ACID IN DILUTE SULPHURIC ACID OF DIFFERENT SPECIFIC GRAVITIES. (URE.)

Specific Gravity.	Real Acid in 100 parts of the Liquid.	Specific Gravity.	Real Acid in 100 parts of the Liquid.	Specific Gravity.	Real Acid in 100 parts of the Liquid.
1·8485	81·54	1·8115	73·39	1·7120	65·23
1·8475	80·72	1·8043	72·57	1·6993	64·42
1·8460	79·90	1·7962	71·75	1·6870	63·60
1·8439	79·09	1·7870	70·94	1·6750	62·78
1·8410	78·28	1·7774	70·12	1·6630	61·97
1·8376	77·46	1·7673	69·31	1·6520	61·15
1·8336	76·65	1·7570	68·49	1·6415	60·34
1·8290	75·83	1·7465	67·68	1·6321	59·52
1·8233	75·02	1·7360	66·86	1·6204	58·71
1·8179	74·20	1·7245	66·05	1·6090	57·89

A TABLE SHOWING THE QUANTITY OF ANHYDROUS ACID IN THE LIQUID NITRIC ACID OF DIFFERENT SPECIFIC GRAVITIES. (URE.)

Specific Gravity.	Real Acid in 100 parts of the Liquid.	Specific Gravity.	Real Acid in 100 parts of the Liquid.	Specific Gravity.	Real Acid in 100 parts of the Liquid.
1·5000	79·700	1·4640	69·339	1·4147	58·978
1·4980	78·903	1·4600	68·542	1·4107	58·181
1·4960	78·106	1·4570	67·745	1·4065	57·384
1·4940	77·309	1·4530	66·948	1·4023	56·587
1·4910	76·512	1·4500	66·155	1·3978	55·790
1·4880	75·715	1·4460	65·354	1·3945	54·993
1·4850	74·918	1·4424	64·557	1·3882	54·196
1·4820	74·121	1·4385	63·760	1·3833	53·399
1·4790	73·324	1·4346	62·963	1·3783	52·602
1·4760	72·527	1·4306	62·166	1·3732	51·805
1·4730	71·730	1·4269	61·369	1·3681	51·068
1·4700	70·933	1·4228	60·572	1·3630	50·211
1·4670	70·136	1·4189	59·775	1·3579	49·414

WEIGHTS AND MEASURES.

Troy, or Apothecaries' Weight.

1 Pound = 12 Ounces. 1 Ounce = 8 Drachms. 1 Drachm = 3 Scruples. 1 Scruple = 20 Grains. (1 Ounce Troy = 480 Grains, or 1 Ounce Avoirdupois *plus* 42·5 grains.)

Avoirdupois Weight.

1 Pound = 16 Ounces. 1 Ounce = 16 Drachms. 1 Drachm = 27·343 grains. (1 Ounce Avoirdupois = 437·5 grains.) (1 Pound Avoirdupois = 7000 Grains, or 1 Pound Troy *plus* 2½ Troy Ounces *plus* 40 grains.)

Imperial Measure.

1 Gallon = 8 Pints. 1 Pint = 20 Ounces. 1 Ounce = 8 Drachms. 1 Drachm = 60 Minims. (A Wine Pint of water measures 16 Ounces, and weighs a Pound.)

An Imperial Gallon of water *weighs* 10 Pounds Avoirdupois, or 70,000 Grains. An Imperial Pint of water *weighs* 1¼ Pound Avoirdupois. A fluid Ounce of water *weighs* 1 Ounce Avoirdupois, or 437·5 Grains. A Drachm of water *weighs* 54·7 Grains.

French Measures of Weight.

1 Kilogramme = 1000 Grammes = something less than 2¼ Pounds Avoirdupois.

1 Gramme = 10 Décigrammes = 100 Centigrammes = 1000 Milligrammes = 15·433 English Grains.

A Gramme of water *measures* 17 English Minims, nearly. 1000 Grammes of water *measure* 35¼ English fluid Ounces.

French Measures of Volume.

1 Litre = 13 Décilitres = 100 Centilitres = 1000 Millilitres = 35¼ English fluid Ounces.

1 Litre = 1 Cubic Décimètre = 1000 Cubic Centimètres.

1 Cubic Centimètre = 17 English Minims.

A Litre of water *weighs* a Kilogramme, or something less than 2¼ Pounds Avoirdupois. A Cubic Centimètre of water *weighs* a Gramme.

INDEX.*

———

* The preparation and properties of the Chemicals used in Photography will be found in the Alphabetical List commencing at page 327.

PRINTED BY

JOHN EDWARD TAYLOR, LITTLE QUEEN STREET,

LINCOLN'S INN FIELDS.

39

ЭЛ $\frac{4}{24}$

194. HARDWICH, T. FREDERICK, *A Manual Of Photographic Chemistry, Including the Practise of the Collodion Process.* 4th edn., 1857, pp. 379. (Incunabula 739).

www.ingramcontent.com/pod-product-compliance
Lightning Source LLC
Chambersburg PA
CBHW071358050326
40689CB00010B/1685